Math Made Nice-n-Easy Books™

 # 4

In This Book:

- ## Complex Numbers
- ## Quadratic Equations
- ## Plane & Solid Geometry
- ## Trigonometry

"MATH MADE NICE-n-EASY # 4" is one in a series of books designed to make the learning of math interesting and fun. For help with additional math topics, see the complete series of "MATH MADE NICE-n-EASY" titles.

Based on U.S. Government Teaching Materials

Research & Education Association
61 Ethel Road West
Piscataway, New Jersey 08854

MATH MADE NICE-N-EASY BOOKS™
BOOK #4

Year 2002 Printing

Printed in the United States of America

Library of Congress Control Number 99-70139

International Standard Book Number 0-87891-203-7

MATH MADE NICE-N-EASY is a trademark of
Research & Education Association, Piscataway, New Jersey 08854

WHAT "MATH MADE NICE-N-EASY" WILL DO FOR YOU

The "Math Made Nice-n-Easy" series simplifies the learning and use of math and lets you see that math is actually interesting and fun. This series of books is for people who have found math scary, but who nevertheless need some understanding of math without having to deal with the complexities found in most math textbooks.

The "Math Made Nice-n-Easy" series of books is useful for students and everyone who needs to acquire a basic understanding of one or more math topics. For this purpose, the series is divided into a number of books which deal with math in an easy-to-follow sequence beginning with basic arithmetic, and extending through pre-algebra, algebra, and calculus. Each topic is described in a way that makes learning and understanding easy.

Almost everyone needs to know at least some math at work, or in a course of study.

For example, almost all college entrance tests and professional exams require solving math problems. Also, almost all occupations (waiters, sales clerks, office people) and all crafts (carpentry, plumbing, electrical) require some ability in math problem solving.

The "Math Made Nice-n-Easy" series helps the reader grasp quickly the fundamentals that are needed in using math. The reader is led by the hand, step-by-step, through the various concepts and how they are used.

By acquiring the ability to use math, the reader is encouraged to further his/her skills and to forget about any initial math fears.

The "Math Made Nice-n-Easy" series includes material originated by U.S. Government research and educational efforts. The research was aimed at devising tutoring and teaching methods for educating government personnel lacking a technical and/or mathematical background. Thanks for these efforts are due the U.S. Bureau of Naval Personnel Training.

Dr. Max Fogiel
Program Director

Contents

Chapter 19
NUMERICAL TRIGONOMETRY

Chapter 20
A REVIEW ON SOLVING TRIANGLES

CHAPTER 15

COMPLEX NUMBERS

In certain calculations in mathematics and related sciences, it is necessary to perform operations with numbers unlike any mentioned thus far in this course. These numbers, unfortunately called "imaginary" numbers by early mathematicians, are quite useful and have a very real meaning in the physical sense. The number system, which consists of ordinary numbers and imaginary numbers, is called the COMPLEX NUMBER system. Complex numbers are composed of a "real" part and an "imaginary" part.

This chapter is designed to explain imaginary numbers and to show how they can be combined with the numbers we already know.

REAL NUMBERS

The concept of number, as has been noted in previous chapters, has developed gradually. At one time the idea of number was limited to positive whole numbers.

The concept was broadened to include positive fractions; numbers that lie between the whole numbers. At first, fractions included only those numbers which could be expressed with terms that were integers. Since any fraction may be considered as a ratio, this gave rise to the term RATIONAL NUMBER, which is defined as any number which can be expressed as the ratio of two integers. (Remember that any whole number is an integer.)

It soon became apparent that these numbers were not enough to complete the positive number range. The ratio, π, of the circumference of a circle to its diameter, did not fit the concept of number thus far advanced, nor did such

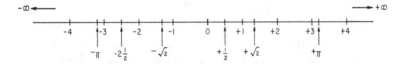

Figure 15-1.—The real number system.

numbers as $\sqrt{2}$ and $\sqrt{3}$. Although decimal values are often assigned to these numbers, they are only approximations. That is, π is not exactly equal to 22/7 or to 3.142. Such numbers are called IRRATIONAL to distinguish them from the other numbers of the system. With rational and irrational numbers, the positive number system includes all the numbers from zero to infinity in a positive direction.

Since the number system was not complete with only positive numbers, the system was expanded to include negative numbers. The idea of negative rational and irrational numbers to minus infinity was an easy extension of the system.

424

Rational and irrational numbers, positive and negative to ± infinity as they have been presented in this course, comprise the REAL NUMBER system. The real number system is pictured in figure 15-1.

OPERATORS

As shown in a previous chapter, the plus sign in an expression such as 5 + 3 can stand for either of two separate things: It indicates the positive number 3, or it indicates that +3 is to be added to 5; that is, it indicates the operation to be performed on +3.

Likewise, in the problem 5 - 3, the minus sign may indicate the negative number -3, in which case the operation would be addition; that is, 5 + (-3). On the other hand, it may indicate the sign of operation, in which case +3 is to be subtracted from 5; that is, 5 - (+3).

Thus, plus and minus signs may indicate positive and negative numbers, or they may indicate operations to be performed.

IMAGINARY NUMBERS

The number line pictured in figure 15-1 represents all positive and negative numbers from plus infinity to minus infinity. However, there is a type of number which does not fit into the picture. Such a number occurs when we try to solve the following equation:

$$x^2 + 4 = 0$$

$$x^2 = -4$$

$$x = \pm \sqrt{-4}$$

425

Notice the distinction between this use of the radical sign and the manner in which it was used in chapter 7. Here, the ± symbol is included with the radical sign to emphasize the fact that two values of x exist. Although both roots exist, only the positive one is usually given. This is in accordance with usual mathematical convention.

The equation

$$x = \pm \sqrt{-4}$$

raises an interesting question:

What number multiplied by itself yields -4? The square of -2 is +4. Likewise, the square of +2 is +4. There is no number in the system of real numbers that is the square root of a negative number. The square root of a negative number came to be called an IMAGINARY NUMBER. When this name was assigned the square roots of negative numbers, it was natural to refer to the other known numbers as the REAL numbers.

IMAGINARY UNIT

To reduce the problem of imaginary numbers to its simplest terms, we proceed as far as possible using ordinary numbers in the solution. Thus, we may write $\sqrt{-4}$ as a product

$$\sqrt{-1.4} = \sqrt{4}\sqrt{-1}$$
$$= \pm 2\sqrt{-1}$$

Likewise,

$$\sqrt{-5} = \sqrt{5}\ \sqrt{-1}$$

Also,

$$3\ \sqrt{-7} = 3\ \sqrt{-7}\ \sqrt{-1}$$

Thus, the problem of giving meaning to the square root of any negative number reduces to that of finding a meaning for $\sqrt{-1}$.

The square root of minus 1 is designated i by mathematicians. When it appears with a coefficient, the symbol i is written last unless the coefficient is in radical form. This convention is illustrated in the following examples:

$$\pm 2\ \sqrt{-1} = \pm 2i$$

$$\sqrt{5}\ \sqrt{-1} = i\ \sqrt{5}$$

$$3\ \sqrt{7}\ \sqrt{-1} = 3i\ \sqrt{7}$$

The symbol i stands for the imaginary unit $\sqrt{-1}$. An imaginary number is any real multiple, positive or negative, of i. For example, -7i, +7i, $i\ \sqrt{15}$, and bi are all imaginary numbers.

In electrical formulas the letter i denotes current. To avoid confusion, electronic technicians use the letter j to indicate $\sqrt{-1}$ and call it "operator j." The name "imaginary" should be thought of as a technical mathematical term of convenience. Such numbers have a very real purpose in the physical sense. Also it can be shown that ordinary mathematical operations such as addition, multiplication, and so forth, may be performed in exactly the same way as for the so-called real numbers.

Practice problems. Express each of the following as some real number times i:

1. $\sqrt{-16}$ 3. $\sqrt{-5}$ 5. $\sqrt{-25}$

2. $2\sqrt{-1}$ 4. $\dfrac{d}{f}\sqrt{-f^2}$ 6. $\sqrt{-\dfrac{9}{16}}$

Answers:

1. $4i$ 3. $i\sqrt{5}$ 5. $5i$

2. $2i$ 4. di 6. $\dfrac{3}{4}i$

Powers of the Imaginary Unit

The following examples illustrate the results of raising the imaginary unit to various powers:

$$i = \sqrt{-1}$$

$$i^2 = \sqrt{-1}\,\sqrt{-1},\ \text{or}\ -1$$

$$i^3 = i^2 i = -1i,\ \text{or}\ -i$$

$$i^4 = i^2 i^2 = -1 \cdot -1 = +1$$

$$i^{-1} = \frac{1}{i} = \frac{i}{i^2} = \frac{i}{-1} = -i$$

We see from these examples that an even power of i is a real number equal to +1 or -1. Every odd power of i is imaginary and equal to i or -i. Thus, all powers of i reduce to one of the following four quantities: $\sqrt{-1}$, -1, $-\sqrt{-1}$, or +1.

GRAPHICAL REPRESENTATION

Figure 15-1 shows the real numbers represented along a straight line, the positive numbers extending from zero to the right for an infinite distance, and the negative numbers extending to the left of zero for an infinite distance. Every point on this line corresponds to a real number, and there are no gaps between them. It follows that there is no possibility of representing imaginary numbers on this line.

Earlier, we noted that certain signs could be used as operators. The plus sign could stand for the operation of addition. The minus sign could stand for the operation of subtraction. Likewise, it is easy to explain the imaginary number i graphically as an operator indicating a certain operation is to be performed on the number of which it is the coefficient.

If we graphically represent the length, n, on the number line pictured in figure 15-2 (A), we start at the point 0 and measure to the right (positive direction) a distance representing n units. If we multiply n by -1, we may represent the result -n by measuring from 0 in a negative direction a distance equal to n units.

Graphically, multiplying a real number by -1 is equivalent to rotating the line that represents the number about the point 0 through 180° so that the new position of n is in the opposite direction and a distance n units from 0. In this case we may think of -1 as the operator that rotates n through two right angles to its new position (fig. 15-2 (B)).

As we have shown, $i^2 = -1$. Therefore, we have really multiplied n by i^2, or i x i. In other words, multiplying by -1 is the same as multiplying by i twice in succession. Logically, if we multiplied n by i just once, the line n would be rotated only half as much as before—that is, through only one right angle, or 90°. The new segment ni would be measured in a direction 90° from the line n. Thus, i is an operator that rotates a number through one right angle. (See fig. 15-3.)

We have shown previously that a positive number may have two real square roots, one positive and one negative. For example, $\sqrt{9} = \pm 3$.

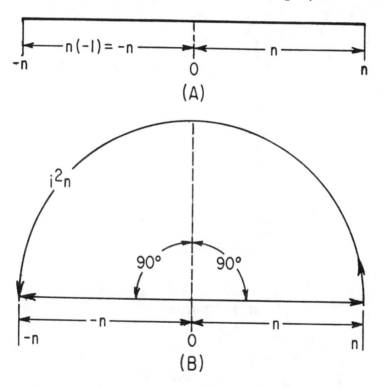

Figure 15-2.—Graphical multiplication by -1 and by operator i^2.

We also saw that an imaginary number may have two roots. For example, $\sqrt{-4}$ is equal to

±2i. When the operator -1 graphically rotates a number, it may do so in a counterclockwise or a clockwise direction. Likewise, the operator i may graphically rotate a number in either direction. This fact gives meaning to numbers such as ±2i. It has been agreed that a number multiplied by +i is to be rotated 90° in a counterclockwise direction. A number multiplied by -i is to be rotated 90° in a clockwise direction.

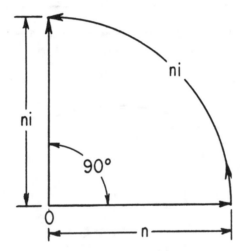

Figure 15-3.—Graphical multiplication by operator i.

In figure 15-4, +2i is represented by rotating the line that represents the positive real number 2 through 90° in a counterclockwise direction. It follows that -2i is represented by rotating the line that represents the positive real number 2 through 90° in a clockwise direction.

In figure 15-5, notice that the idea of i as an operator agrees with the concept advanced con-

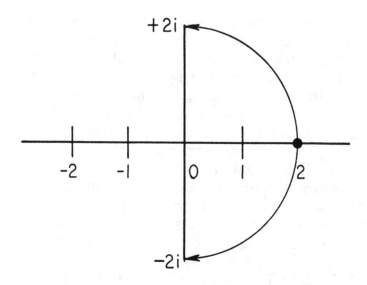

Figure 15-4.—Graphical representation of ±2i.

cerning the powers of i. Thus, i rotates a number through 90°; i^2 or -1 rotates the number through 180°, and the number is real and negative; i^3 rotates the number through 270°, which has the same effect as -1; and i^4 rotates the number through 360°, and the number is once again positive and real.

THE COMPLEX PLANE

All imaginary numbers may be represented graphically along a line extending through zero and perpendicular to the line representing the real numbers. This line may be considered infinite in both the positive and negative directions, and all multiples of i may be represented on it. This graph is similar to the rectangular coordinate system studied earlier.

In this system, the vertical or y axis is called the axis of imaginaries, and the horizontal or x axis is called the axis of reals. In the rectangular coordinate system, real numbers are laid off on both the x and y axes and the plane on which the axes lie is called the real plane. When the y axis is the axis of imaginaries, the plane determined by the x and y axes is called the COMPLEX PLANE (fig. 15-6).

In any system of numbers a unit is necessary for counting. Along the real axis, the unit is the number 1. As shown in figure 15-6, along the imaginary axis the unit is i. Numbers that lie along the imaginary axis are called PURE IMAGINARIES. They will always be some multiple of i, the imaginary unit. The numbers $5i$, $3i\sqrt{2}$, and $\sqrt{-7}$ are examples of pure imaginaries.

NUMBERS IN THE COMPLEX PLANE

All numbers in the complex plane are complex numbers, including reals and pure imaginaries. However, since the reals and imaginaries have the special property of being located on the axes, they are usually identified by their distinguishing names.

The term complex number has been defined as the indicated sum or difference of a real number and an imaginary number.

For example, $3 + 5\sqrt{-1}$ or $3 + 5i$, $2 - 6i$, and $-2 + \sqrt{-5}$ are complex numbers. In the complex number $7 - i\sqrt{2}$, 7 is the real part and $-1\sqrt{2}$ is the imaginary part.

All complex numbers correspond to the gen-

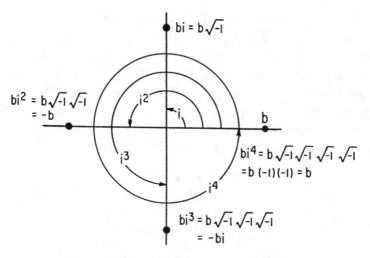

Figure 15-5.—Operation with powers of i.

eral form a + bi, where a and b are real numbers. When a has the value 0, the real term disappears and the complex number becomes a pure imaginary. When b has the value of 0, the imaginary term disappears and the complex number becomes a real number. Thus, 4 may

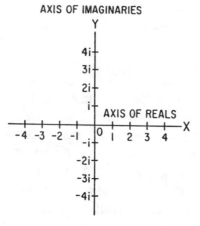

Figure 15-6.—The complex plane.

be thought of as 4 + 0i, and 3i may be considered 0 + 3i. From this we may reason that the real number and the pure imaginary number are special cases of the complex number. Consequently, the complex number may be thought of as the most general form of a number and can be construed to include all the numbers of algebra as shown in the chart in figure 15-7.

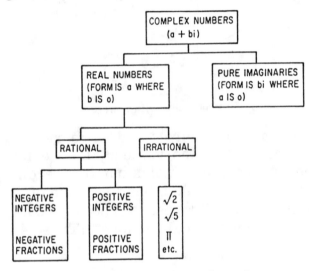

Figure 15-7.—The complex number system.

Plotting Complex Numbers

Complex numbers may easily be plotted in the complex plane. Pure imaginaries are plotted along the vertical axis, the axis of imaginaries, and real numbers are plotted along the horizontal axis, the axis of reals. It follows that other points in the complex plane must represent numbers that are part real and part imaginary; in other words, complex numbers. If we wish to plot the point 3 + 2i, we note that the number is made up of the real number 3 and the imaginary number 2i. Thus, as in figure 15-8, we measure along the real axis in a

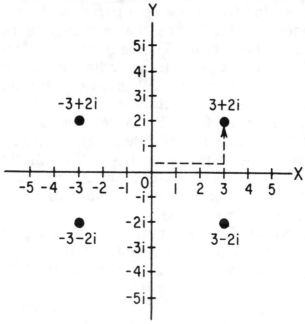

Figure 15-8.—Plotting complex numbers.

positive direction. At point (3, 0) on the real axis we turn through one right angle and measure 2 units up and parallel to the imaginary axis. Likewise, the number -3 + 2i is 3 units to the left and up 2 units; the number 3 - 2i is 3 units to the right and down 2 units; and the number -3 -2i is 3 units to the left and down 2 units.

Complex Numbers as Vectors

A vector is a directed line segment. A complex number represents a vector expressed in the RECTANGULAR FORM. For example, the complex number 6 + 8i in figure 15-9 may be considered as representing either the point P or the line OP. The real parts of the complex number (6 and 8) are the rectangular components of the vector. The real parts are the legs of the right triangle (sides adjacent to the right angle), and the vector OP is its hypotenuse (side opposite the right angle). If we merely wish to indicate the vector OP, we may do so by writing the complex number that represents it along the segment as in figure 15-9. This method not only fixes the position of point P, but also shows what part of the vector is imaginary (PA) and what part is real (OA).

If we wish to indicate a number that shows the actual length of the vector OP, it is necessary to solve the right triangle OAP for its hypotenuse. This may be accomplished by taking the square root of the sum of the squares of the legs of the triangle, which in this case are the real numbers, 6 and 8. thus,

$$OP = \sqrt{6^2 + 8^2}$$
$$= \sqrt{100}$$
$$= 10$$

However, since a vector has direction as well as magnitude, we must also show the direction of the segment; otherwise the segment OP could radiate in any direction on the complex plane from point 0. The expression $10\underline{/53.1°}$ indicates that the vector OP has been rotated counterclockwise from the initial posi-

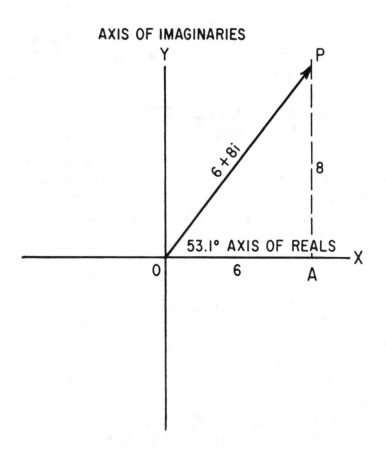

Figure 15-9.—A complex number
shown as a vector.

tion through an angle of 53.1°. (The initial position in a line extending from the origin to the
right along OX.) This method of expressing the
vector quantity is called the POLAR FORM.
The number represents the magnitude of the
quantity, and the angle represents the position
of the vector with respect to the horizontal reference, OX.   Positive angles represent counterclockwise rotation of the vector, and negative angles represent clockwise rotation.   The

polar form is generally simpler for multiplication and division, but its use requires a knowledge of trigonometry.

ADDITION AND SUBTRACTION OF COMPLEX NUMBERS

Pure imaginaries are added and subtracted in the same way as any other algebraic quantities. The coefficients of similar terms are added or subtracted algebraically, as follows:

$$4i + 3i = 7i$$

$$4i - 3i = i$$

$$4i - (-3i) = 7i$$

Likewise, complex numbers in the rectangular form are combined like any other algebraic polynomials. Add or subtract the coefficients of similar terms algebraically. If parentheses enclose the numbers, first remove the parentheses. Next, place the real parts together and the imaginary parts together. Collect terms. As examples, consider the following:

1. $(2 - 3i) + (5 + 4i) = 2 - 3i + 5 + 4i$
$$= 2 + 5 - 3i + 4i$$
$$= 7 + i$$

2. $(2 - j3) - (5 + j4) = 2 - j3 - 5 - j4$
$$= 2 - 5 - j3 - j4$$
$$= -3 - j7$$

In example 2, notice that the convention for writing operator j (the electronics form of the imaginary unit) with numerical coefficients is to place j first.

If the complex numbers are placed one under the other, the results of addition and subtraction appear as follows:

ADDITION	SUBTRACTION
$3 + 4\sqrt{-1}$	$a + jb$
$\underline{2 - 7\sqrt{-1}}$	$\underline{[\leftarrow c + jd}$
$5 - 3\sqrt{-1}$	$(a - c) + j(b - d)$

Practice problems. Add or subtract as indicated, in the following problems:

1. $(3a + 4i) + (0 - 2i)$
2. $(3 + 2i) + (-3 + 3i)$
3. $(a + bi) + (c + di)$
4. $(1 + 2\sqrt{-1}) + (-2 - 2\sqrt{-1})$
5. $(-5 + 3i) - (4 - 2i)$
6. $(a + bi) - (-c + di)$

Answers:

1. $3a + 2i$ 4. -1
2. $5i$ 5. $-9 + 5i$
3. $a + c + (b + d)i$ 6. $a + c + (b - d)i$

MULTIPLICATION OF COMPLEX NUMBERS

Generally, the rules for the multiplication of complex numbers and pure imaginaries are the same as for other algebraic quantities. However, there is one exception that should be noted: The rule for multiplying numbers under radical signs does not apply to TWO NEGATIVE numbers. When at least one of two radicands is positive, the radicands can be multiplied immediately, as in the following examples:

$$\sqrt{2}\,\sqrt{3} = \sqrt{6}$$
$$\sqrt{2}\,\sqrt{-3} = \sqrt{-6}$$

When both radicands are negative, however, as in $\sqrt{-2}\,\sqrt{-3}$, an inconsistent result is obtained if we multiply both numbers under the radical signs immediately. To get the correct result, express the imaginary numbers first in terms of i, as follows:

$$\sqrt{-2}\,\sqrt{-3} = i\,\sqrt{2} \cdot i\,\sqrt{3}$$
$$= i^2\sqrt{2}\,\sqrt{3}$$
$$= i^2\sqrt{6}$$
$$= (-1)\,\sqrt{6} = -\sqrt{6}$$

Multiplying complex numbers is equivalent to multiplying binomials in the manner explained previously. After the multiplication is performed, simplify the powers of i as in the following examples:

1. $4 - i$

$\underline{3 + i}$

$12 - 3i$

$\underline{ + 4i - i^2}$

$12 + i - i^2 = 12 + i - (-1)$

$$= 13 + i$$

2. $(-6 + 5\sqrt{-7})(8 - 2\sqrt{-7})$

$= (-6 + 5i\sqrt{7})(8 - 2i\sqrt{7})$

$= -48 + 40i\sqrt{7} + 12i\sqrt{7} - 10(7)i^2$

$= -48 + 52i\sqrt{7} + 70$

$= 22 + 52i\sqrt{7}$

Practice problems. Perform the indicated operations:

1. $\sqrt{-9}\ \sqrt{-16}$
2. $\sqrt{-2}\ \sqrt{18}$
3. $\sqrt{-9}\ \sqrt{-4}$
4. $a\ \sqrt{-ba}\ \cdot\ \sqrt{-b}$
5. $(2 + 5i)(3 - 2i)$
6. $(a + \sqrt{-b})(a - \sqrt{-b})$
7. $(-2 + \sqrt{-4})(-1 + \sqrt{-4})$
8. $(8 - \sqrt{-7})(6 + \sqrt{-7})$

Answers:

1. -12   5. 16 + 11i

2. 6i   6. $a^2 + b$

3. 6   7. -2 - 6i

4. -ab $\sqrt{a}$ 8. 55 + 2i $\sqrt{7}$

CONJUGATES AND SPECIAL PRODUCTS

Two complex numbers that are alike except for the sign of their imaginary parts are called CONJUGATE COMPLEX NUMBERS. For example, 3 + 5i and 3 - 5i are conjugates. Either number is the conjugate of the other.

If one complex number is known, the conjugate can be obtained immediately by changing the sign of the imaginary part. The conjugate of -8 + $\sqrt{-10}$ is -8 - $\sqrt{-10}$. The conjugate of - $\sqrt{-6}$ is $\sqrt{-6}$.

The sum of two conjugate complex numbers is a real number, as illustrated by the following:

1. (3 + j5) + (3 - j5) = 2(3) = 6

2. $\left(-\dfrac{1}{2} + \dfrac{\sqrt{-3}}{2}\right) + \left(-\dfrac{1}{2} - \dfrac{\sqrt{-3}}{2}\right)$

$$= -\dfrac{1}{2} + \dfrac{\sqrt{3}}{2}i - \dfrac{1}{2} - \dfrac{\sqrt{3}}{2}i$$

$$= -\dfrac{1}{2} + \left(-\dfrac{1}{2}\right)$$

$$= -1$$

443

Product of Two Conjugates

The product of two conjugate complex numbers is a real number. Multiplying two conjugates is equivalent to finding the product of the sum and difference of two numbers.

Consider the following examples:

1. $(3 + j5)(3 - j5) = 3^2 - (j5)^2$

$$= 9 - 25(-1)$$
$$= 9 + 25$$
$$= 34$$

2. $\left(-\dfrac{1}{2} + \dfrac{\sqrt{3}}{2}i\right)\left(-\dfrac{1}{2} - \dfrac{\sqrt{3}}{2}i\right) = \left(-\dfrac{1}{2}\right)^2 - \left(\dfrac{\sqrt{3}}{2}i\right)^2$

$$= \dfrac{1}{4} - \left[\dfrac{3}{4}(-1)\right]$$
$$= \dfrac{1}{4} + \dfrac{3}{4}$$
$$= 1$$

Squaring a Complex Number

Squaring a complex number is equivalent to raising a binomial to the second power. For example:

$(-6 - \sqrt{-25})^2 = (-6 - j5)^2$

$$= [(-1) \cdot (6 + j5)]^2$$
$$= (-1)^2 \cdot (6^2 + j60 + j^2 25)$$
$$= 36 + j60 - 25$$
$$= 11 + j60$$

DIVISION OF COMPLEX NUMBERS

When dividing by a pure imaginary, the denominator may be rationalized and the problem thus simplified by multiplying both numerator and denominator by the denominator. Thus,

$$\frac{12}{\sqrt{-2}} = \frac{12}{i\sqrt{2}} \cdot \frac{i\sqrt{2}}{i\sqrt{2}}$$

$$= \frac{12i\sqrt{2}}{2i^2}$$

$$= \frac{6i\sqrt{2}}{-1}$$

$$= -6i\sqrt{2}$$

Division of complex numbers can be accomplished by multiplying the numerator and denominator by the number that is the conjugate of the denominator. This process is similar to the process of rationalizing a denominator in the case of real numbers that are irrational.

As an example, consider

$$\frac{5 - 2i}{3 + i}$$

The denominator is $3 + i$. Its conjugate is $3 - i$. Multiplying numerator and denominator by $3 - i$ gives

445

$$\frac{5 - 2i}{3 + i} \cdot \frac{3 - i}{3 - i} = \frac{15 - 11i + 2i^2}{9 - i^2}$$

$$= \frac{15 - 11i - 2}{9 + 1}$$

$$= \frac{13 - 11i}{10}$$

$$= \frac{13}{10} - \frac{11}{10}i$$

Practice problems. Rationalize the denominators and simplify:

1. $\dfrac{2 \sqrt{-1}}{4 + 2 \sqrt{-1}}$

2. $\dfrac{-2 + 4i}{-1 + 4i}$

3. $\dfrac{3 + \sqrt{-2}}{3 - \sqrt{-2}}$

4. $\dfrac{3}{1 - i \sqrt{3}}$

5. $\dfrac{1 - i}{2 - i}$

6. $\dfrac{8}{2 + \sqrt{-2}}$

Answers:

1. $\dfrac{2i + 1}{5}$

2. $\dfrac{18 + 4i}{17}$

3. $\dfrac{7 + 6i \sqrt{2}}{11}$

4. $\dfrac{3}{4} + \dfrac{3}{4} i \sqrt{3}$

5. $\dfrac{3 - i}{5}$

6. $\dfrac{8 - 4i \sqrt{2}}{3}$

CHAPTER 16

QUADRATIC EQUATIONS IN ONE VARIABLE

The degree of an equation in one variable is the exponent of the highest power to which the variable is raised in that equation. A second-degree equation in one variable is one in which the variable is raised to the second power. A second-degree equation is often called a QUADRATIC EQUATION. The word quadratic is derived from the Latin word quadratus, which means "squared." In a quadratic equation the term of highest degree is the squared term. For example, the following are quadratic equations:

$$x^2 + 3x + 4 = 0$$

$$3m + 4m^2 = 6$$

The terms of degree lower than the second may or may not be present. The possible terms of lower degree than the squared term in a quadratic equation are the first-degree term and the constant term. In the equation

$$3x^2 - 8x - 5 = 0$$

-5 is the coefficient of x^0. If we wished to emphasize the powers of x in this equation, we could write the equation in the form

$$3x^2 - 8x^1 - 5x^0 = 0$$

Examples of quadratic equations in which either the first-degree term or the constant term is missing are:

1. $4x^2 = 16$
2. $y^2 + 16y = 0$
3. $e^2 + 12 = 0$

GENERAL FORM OF A QUADRATIC EQUATION

Any quadratic equation can be arranged in the general form:

$$ax^2 + bx + c = 0$$

If it has more than three terms, some of them will be alike and can be combined, after which the final form will have at most three terms. For example,

$$2x^2 + 3 + 5x - 1 + x^2 = 4 - x^2 - 2x - 3$$

reduces to the simpler form

$$4x^2 + 7x + 1 = 0$$

In this form, it is easy to see that a, the coefficient of x^2, is 4; b, the coefficient of x, is 7; and c, the constant term, is 1.

448

Sometimes the coefficients of the terms of a quadratic appear as negative numbers, as follows:

$$2x^2 - 3x - 5 = 0$$

This equation can be rewritten in such a way that the connecting signs are all positive, as in the general form. This is illustrated as follows:

$$2x^2 + (-3)x + (-5) = 0$$

In this form, the value of a is seen to be 2, b is -3, and c is -5.

An equation of the form

$$x^2 + 2 = 0$$

has no x term. This can be considered as a case in which a is 1 (coefficient of x^2 understood to be 1), b is 0, and c is 2. For the purpose of emphasizing the values of a, b, and c with reference to the general form, this equation can be written

$$x^2 + 0x + 2 = 0$$

The coefficient of x^2 can never be 0; if it were 0, the equation would not be a quadratic. If the coefficients of x and x^0 are 0, then those terms do not normally appear. To say that the coefficient of x^0 is 0 is the same as saying that the constant term is 0 or is missing.

A ROOT of an equation in one variable is a value of the variable that satisfied the equation.

Every equation in one variable, with constants as coefficients and positive integers as exponents, has as many roots as the exponent of the highest power. In other words, the number of roots is the same as the degree of the equation.

A fourth-degree equation has four roots, a cubic (third-degree) equation has three roots, a quadratic equation has two roots, and a linear equation has one root.

As an example, 6 and -1 are roots of the quadratic equation

$$x^2 - 5x - 6 = 0$$

This can be verified by substituting these values into the equation and noting that an identity results in each case.

Substituting x = 6 gives

$$6^2 - 5(6) - 6 = 0$$
$$36 - 36 = 0$$
$$0 = 0$$

Substituting x = -1 gives

$$(-1)^2 - 5(-1) - 6 = 0$$
$$1 + 5 - 6 = 0$$
$$6 - 6 = 0$$
$$0 = 0$$

Several methods of finding the roots of quadratic equations (SOLVING) are possible. The most common methods are solution by FAC-

TORING and solution by the QUADRATIC FOR-
MULA. Less commonly used methods of solution are accomplished by completing the square
and by graphing.

SOLUTION BY FACTORING

The equation $x^2 - 36 = 0$ is a pure quadratic
equation. There are two numbers which, when
substituted for x, will satisfy the equation as
follows:

$$(+6)^2 - 36 = 0$$
$$36 - 36 = 0$$

also

$$(-6)^2 - 36 = 0$$
$$36 - 36 = 0$$

Thus, +6 and -6 are roots of the equation

$$x^2 - 36 = 0$$

The most direct way to solve a pure quadratic (one in which no x term appears and the
constant term is a perfect square) involves re-
writing with the constant term in the right
member, as follows:

$$x^2 = 36$$

Taking square roots on both sides, we have

$$x = \pm 6$$

The reason for expressing the solution as both plus and minus 6 is found in the fact that both +6 and -6, when squared, produce 36.

The equation

$$x^2 - 36 = 0$$

can also be solved by factoring, as follows:

$$x^2 - 36 = 0$$
$$(x + 6)(x - 6) = 0$$

We now have the product of two factors equal to zero. According to the zero factor law, if a product is zero, then one or more of its factors is zero. Therefore, at least one of the factors must be zero, and it makes no difference which one. We are free to set first one factor and then the other factor equal to zero. In so doing we derive two solutions or roots of the equation.

If $x + 6$ is the factor whose value is 0, then we have

$$x + 6 = 0$$
$$x = -6$$

If $x - 6$ is the zero factor, we have

$$x - 6 = 0$$
$$x = 6$$

When a three-term quadratic is put into simplest form, it is customary to place all the terms on the left side of the equality sign with the squared term first, the first-degree term next, and the constant term last, as in

$$9x^2 - 2x + 7 = 0$$

If the trinomial in the left member is readily factorable, the equation can be solved quickly by separating the trinominal into factors. Consider the equation

$$3x^2 - x - 2 = 0$$

By factoring the trinominal, the equation becomes

$$(3x + 2)(x - 1) = 0$$

Once again we have two factors, the product of which is 0. This means that one or the other of them (or both) must have the value 0. If the zero factor is $3x + 2$, we have

$$3x + 2 = 0$$
$$3x = -2$$
$$x = -\frac{2}{3}$$

If the zero factor is $x - 1$, we have

$$x - 1 = 0$$
$$x = 1$$

Substituting first $x = 1$ and then $x = -\frac{2}{3}$ in the original equation, we see that both roots satisfy it. Thus,

$$3(1)^2 - (1) - 2 = 0$$

$$3 - 1 - 2 = 0$$

$$0 = 0$$

$$3\left[-\frac{2}{3}\right]^2 - \left[-\frac{2}{3}\right] - 2 = 0$$

$$\frac{4}{3} + \frac{2}{3} - 2 = 0$$

$$0 = 0$$

In summation, when a quadratic may be readily factored, the process for finding its roots is as follows:

1. Arrange the equation in the order of the descending powers of the variable so that all the terms appear in the left member and zero appears in the right.

2. Factor the left member of the equation.

3. Set each factor containing the variable equal to zero and solve the resulting equations.

4. Check by substituting each of the derived roots in the original equation.

EXAMPLE: Solve the equation $x^2 - 4x = 12$ for x.

1. $x^2 - 4x - 12 = 0$

2. $(x - 6)(x + 2) = 0$

3. $x - 6 = 0$ $x + 2 = 0$

     $x = 6$ $x = -2$

4. $(6)^2 - 4(6) = 12$ (x = 6)

$$36 - 24 = 12$$

$$12 = 12$$

$$(-2)^2 - 4(-2) = 12 \quad (x = -2)$$

$$4 + 8 = 12$$

$$12 = 12$$

Practice problems. Solve the following equations by factoring:

1. $x^2 + 10x - 24 = 0$ 4. $7y^2 - 19y - 6 = 0$

2. $a^2 - a - 56 = 0$ 5. $m^2 - 4m = 96$

3. $y^2 - 2y = 63$

Answers:

1. x = -12   4. y = 3

    x = 2 $y = -\dfrac{2}{7}$

2. a = 8   5. m = -8

    a = -7   m = 12

3. y = -7

    y = 9

SOLUTION BY COMPLETING THE SQUARE

When a quadratic cannot be solved by factoring, or the factors are not readily seen, another method of finding the roots is needed. A method that may always be used for quadratics in one variable involves perfect square trinomials. These, we recall, are trinomials whose factors are identical. For example,

$$x^2 - 10x + 25 = (x - 5)(x - 5) = (x - 5)^2$$

Recall that in squaring a binomial, the third term of the resulting perfect square trinomial is always the square of the second term of the binomial. The coefficient of the middle term of the trinomial is always twice the second term of the binomial. For example, when $(x + 4)$ is squared, we have

$$
\begin{array}{r}
x + 4 \\
x + 4 \\
\hline
x^2 + 4x \\
+ 4x + 16 \\
\hline
x^2 + 8x + 16
\end{array}
$$

Hence if both the second- and first-degree terms of a perfect square trinomial are known, the third may be written by squaring one-half the coefficient of the first-degree term.

Essentially, in completing the square, certain quantities are added to one member and subtracted from the other, and the equation is so arranged that the left member is a perfect square trinomial. The square roots of both members may then be taken, and the subsequent equalities may be solved for the variable.

For example,

456

$$x^2 + 5x - \frac{11}{4} = 0$$

cannot be readily factored. To solve for x by completing the square, we proceed as follows:

1. Leave only the second- and first-degree terms in the left member.

$$x^2 + 5x = \frac{11}{4}$$

(If the coefficient of x^2 is not 1, divide through by the coefficient of x^2.)

2. Complete the square by adding to both members the square of half the coefficient of the x term. In this example, one-half of the coefficient of the x term is $\frac{5}{2}$, and the square of $\frac{5}{2}$ is $\frac{25}{4}$. Thus,

$$x^2 + 5x + \frac{25}{4} = \frac{11}{4} + \frac{25}{4}$$

3. Factor the left member and simplify the right member.

$$\left(x + \frac{5}{2}\right)^2 = 9$$

4. Take the square root of both members.

$$\sqrt{\left(x + \frac{5}{2}\right)^2} = \sqrt{9}$$

$$x + \frac{5}{2} = \pm 3$$

Remember that, in taking square roots on both sides of an equation, we must allow for the fact that two roots exist in every second-degree equation. Thus we designate both the plus and the minus root of 9 in this example.

5. Solve the resulting equations.

$$x + \frac{5}{2} = 3 \qquad\qquad x + \frac{5}{2} = -3$$

$$x = \frac{6}{2} - \frac{5}{2} \qquad\qquad x = -\frac{6}{2} - \frac{5}{2}$$

$$x = \frac{1}{2} \qquad\qquad x = -\frac{11}{2}$$

6. Check the results.

$$\left(\frac{1}{2}\right)^2 + \frac{5}{2} - \frac{11}{4} = 0$$

$$\frac{5}{2} - \frac{10}{4} = 0$$

$$0 = 0$$

$$\left(-\frac{11}{2}\right)^2 + (5)\left(-\frac{11}{2}\right) - \frac{11}{4} = 0$$

$$\frac{121}{4} - \frac{55}{2} - \frac{11}{4} = 0$$

$$\frac{110}{4} - \frac{55}{2} = 0$$

$$0 = 0$$

The process of completing the square may always be used to solve a quadratic equation.

However, since this process may become complicated in more complex equations, a formula based on completing the square has been developed in which known quantities may be substituted in order to derive the roots of the quadratic equation. This formula is explained in the following paragraphs.

SOLUTION BY THE QUADRATIC FORMULA

The quadratic formula is derived by applying the process of completing the square to solve for x in the general form of the quadratic equation, $ax^2 + bx + c = 0$. Remember that, the general form represents every possible quadratic equation. Thus, if we can solve this equation for x, the solution will be in terms of a, b, and c. To solve this equation for x by completing the square, we proceed as follows:

1. Subtract the constant term, c, from both members.

$$ax^2 + bx + c = 0$$
$$ax^2 + bx = -c$$

2. Divide all terms by a so that the coefficient of the x^2 term becomes unity.

$$x^2 + \frac{b}{a}x = -\frac{c}{a}$$

3. Add the square of one-half the coefficient of the x term, $\frac{b}{a}$, to both members.

Square $\dfrac{b}{2a}$: $\left(\dfrac{b}{2a}\right)^2 = \dfrac{b^2}{4a^2}$

Add: $x + \dfrac{b}{a}x + \dfrac{b^2}{4a^2} = \dfrac{b^2}{4a^2} - \dfrac{c}{a}$

4. Factor the left member and simplify the right member.

$$\left(x + \dfrac{b}{2a}\right)^2 = \dfrac{b^2 - 4ac}{4a^2}$$

5. Take the square root of both members.

$$x + \dfrac{b}{2a} = \pm \dfrac{\sqrt{b^2 - 4ac}}{2a}$$

6. Solve for x.

$$x = -\dfrac{b}{2a} \pm \dfrac{\sqrt{b^2 - 4ac}}{2a}$$

$$= \dfrac{-b \pm \sqrt{b^2 - 4ac}}{2a}$$

Thus, we have solved the equation representing every quadratic for its unknown in terms of its constants a, b, and c. Hence, in a given quadratic we need only substitute in the expression

$$\dfrac{-b \pm \sqrt{b^2 - 4ac}}{2a}$$

the values of a, b, and c, as they appear in the particular equation, to derive the roots of that equation. This expression is called the QUAD-

RATIC FORMULA. The general quadratic equation, $ax^2 + bx + c = 0$, and the quadratic formula should be memorized. Then, when a quadratic cannot be solved quickly by factoring, it may be solved at once by the formula.

EXAMPLE: Use the quadratic formula to solve the equation

$$x^2 + 30 - 11x = 0.$$

SOLUTION:

1. Set up the equation in standard form.

$$x^2 - 11x + 30 = 0$$

Then a (coefficient of x^2) = 1

　　 b (coefficient of x) = -11

　　 c (the constant term) = 30

2. Substituting,

$$x = \frac{-b \pm \sqrt{b^2 - 4ac}}{2a}$$

$$= \frac{-(-11) \pm \sqrt{(-11)^2 - 4(1)(30)}}{2(1)}$$

$$= \frac{11 \pm \sqrt{121 - 120}}{2}$$

$$= \frac{11 \pm 1}{2} = 6 \text{ or } 5$$

3. Checking:

When

$$x = 6,$$
$$(6)^2 - 11(6) + 30 = 0$$
$$36 - 66 + 30 = 0$$
$$0 = 0$$

When

$$x = 5,$$
$$(5)^2 - 11(5) + 30 = 0$$
$$25 - 55 + 30 = 0$$
$$0 = 0$$

EXAMPLE: Find the roots of

$$2x^2 - 3x - 1 = 0$$

Here, a = 2, b = -3, and c = -1.

Substituting into the quadratic formula gives

$$x = \frac{-(-3) \pm \sqrt{(-3)^2 - 4(2)(-1)}}{2(2)}$$

$$= \frac{3 \pm \sqrt{9 + 8}}{4}$$

$$= \frac{3 \pm \sqrt{17}}{4}$$

The two roots are

$$x = \frac{3}{4} + \frac{1}{4} \sqrt{17} \text{ and } x = \frac{3}{4} - \frac{1}{4} \sqrt{17}$$

These roots are irrational numbers, since the radicals cannot be removed.

If the decimal values of the roots are desired, the value of the square root of 17 can be taken from appendix I of this course. Substituting $\sqrt{17} = 4.1231$ and simplifying gives

$$x_1 = \frac{3 + 4.1231}{4} \quad \text{and} \quad x_2 = \frac{3 - 4.1231}{4}$$

$$x_1 = \frac{7.1231}{4} \qquad\qquad x_2 = \frac{-1.1231}{4}$$

$$x_1 = 1.781 \qquad\qquad x_2 = -0.281$$

In decimal form, the roots of $2x^2 - 3x - 1 = 0$ to the nearest tenth are 1.8 and -0.3.

Notice that the subscripts, 1 and 2, are used to distinguish between the two roots of the equation. The three roots of a cubic equation in x might be designated x_1, x_2, and x_3. Sometimes the letter r is used for root. Using r, the roots of a cubic equation could be labeled r_1, r_2, and r_3.

Checking:

When
$$x_1 = \frac{3 + \sqrt{17}}{4}$$

$$2x^2 - 3x - 1 = 0$$

then

$$2\left(\frac{3 + \sqrt{17}}{4}\right)^2 - 3\left(\frac{3 + \sqrt{17}}{4}\right) - 1 = 0$$

$$\frac{(3 + \sqrt{17})^2}{8} - \frac{9 + 3\sqrt{17}}{4} - 1 = 0$$

$$\frac{9 + 6\sqrt{17} + 17 - 18 - 6\sqrt{17} - 8}{8} = 0$$

$$0 = 0$$

463

When

$$x_2 = \frac{3 - \sqrt{17}}{4}$$

then

$$2\left(\frac{3 - \sqrt{17}}{4}\right)^2 - 3\left(\frac{3 - \sqrt{17}}{4}\right) - 1 = 0$$

$$\frac{9 - 6\sqrt{17} + 17}{8} - \frac{9 - 3\sqrt{17}}{4} - 1 = 0$$

Multiplying both members of the equation by 8, the LCD, we have

$$8\left(\frac{9 - 6\sqrt{17} + 17}{8}\right) - 8\left(\frac{9 - 3\sqrt{17}}{4}\right) - 8(1) = 0$$

$$9 - 6\sqrt{17} + 17 - 2(9 - 3\sqrt{17}) - 8 = 0$$

$$9 - 6\sqrt{17} + 17 - 18 + 6\sqrt{17} - 8 = 0$$

$$0 = 0$$

Practice problems. Use the quadratic formula to find the roots of the following equations:

1. $3x^2 - 20 - 7x = 0$

2. $4x^2 - 3x - 5 = 0$

3. $15x^2 - 22x - 5 = 0$

4. $x^2 + 7x = 8$

Answers:

1. $x_1 = 4$

   $x_2 = -\dfrac{5}{3}$

2. $x_1 = \dfrac{3 + \sqrt{89}}{8}$

   $x_2 = \dfrac{3 - \sqrt{89}}{8}$

3. $x_1 = \dfrac{5}{3}$ 4. $x_1 = 1$

   $x_2 = -\dfrac{1}{5}$ $x_2 = -8$

GRAPHICAL SOLUTION

A fourth method of solving a quadratic equation is by means of graphing. In graphing linear equations using both axes as reference, we recall that an independent variable, x, and a dependent variable, y, were needed. The coordinates of points on the graph of the equation were designated (x, y).

Since the quadratics we are considering contain only one variable, as in the equation

$$x^2 - 8x + 12 = 0$$

we cannot plot values for the equations in the present form using both x and y axes. A dependent variable, y, is necessary.

If we think of the expression

$$x^2 - 8x + 12$$

as a function, then this function can be considered to have many possible numerical values, depending on what value we assign to x. The particular value or values of x which cause the value of the function to be 0 are solutions for the equation

$$x^2 - 8x + 12 = 0$$

For convenience, we may choose to let y

represent the function

$$x^2 - 8x + 12$$

If numerical values are now assigned to x, the corresponding values of y may be calculated. When these pairs of corresponding values of x and y are tabulated, the resulting table provides the information necessary for plotting a graph of the function.

EXAMPLE: Graph the equation

$$x^2 + 2x - 8 = 0$$

and from the graph write the roots of the equation.

SOLUTION:

1. Let $y = x^2 + 2x - 8$.

2. Make a table of the y values corresponding to the value assigned x, as shown in table 16-1.

Table 16-1.--Tabulation of x and y values
for the function $y = x^2 + 2x - 8$.

if x = -----	-5	-4	-3	-2	-1	0	1	2	3
then y ---	7	0	-5	-8	-9	-8	-5	0	7

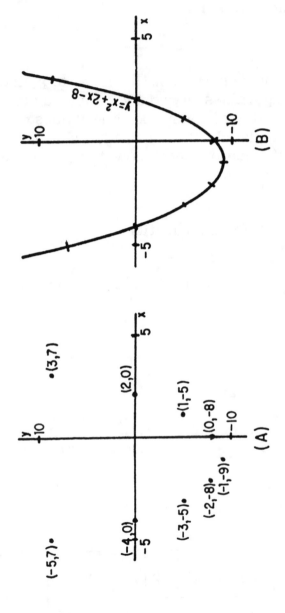

Figure 16-1.—Graph of the equation $y = x^2 + 2x - 8$.

(A) Points plotted;

(B) curve drawn through plotted points.

3. Plot the pairs of x and y values that appear in the table as coordinates of points on a rectangular coordinate system as in figure 16-1 (A).

4. Draw a smooth curve through these points, as shown in figure 16-1 (B).

Notice that this curve crosses the X axis in two places. We also recall that, for any point on the X axis, the y coordinate is zero. Thus, in the figure we see that when y is zero, x is -4 or +2. When y is zero, furthermore, we have the original equation,

$$x^2 + 2x - 8 = 0$$

Thus, the values of x at these points where the graph of the equation crosses the X axis (x = -4 or +2) are solutions to the original equation. We may check these results by solving the equation algebraically. Thus,

$$x^2 + 2x - 8 = 0$$

$$(x + 4)(x - 2) = 0$$

$$x_1 + 4 = 0 \qquad x_2 - 2 = 0$$

$$x_1 = -4 \qquad x_2 = 2$$

Check:

$$(-4)^2 + 2(-4) - 8 = 0 \qquad (2)^2 + 2(2) - 8 = 0$$

$$16 - 8 - 8 = 0 \qquad 4 + 4 - 8 = 0$$

$$0 = 0 \qquad 0 = 0$$

The curve in figure 16-1 (B) is called a PARABOLA. Every quadratic of the form

468

$ax^2 + bx + c = y$ will have a graph of this general shape. The curve will open downward if a is negative, and upward if a is positive.

Graphing provides a fourth method of finding the roots of a quadratic in one variable. When the equation is graphed, the roots will be the X intercepts (those values of x where the curve crosses the X axis). The X intercepts are the points at which y is 0.

Practice problems. Graph the following quadratic equations and read the roots of each equation from its graph

1. $x^2 - 4x - 8 = 0$
2. $6x - 5 - x^2 = 0$

Answers:

1. See figure 16-2. $x = 5.5; x = -1.5$
2. See figure 16-3. $x = 1; x = 5$

MAXIMUM AND MINIMUM POINTS

It will be seen from the graphs of quadratics in one variable that a parabola has a maximum or minimum value, depending on whether the curve opens upward or downward. Thus, when a is negative the curve passes through a maximum value; and when a is positive, the curve passes through a minimum value. Often these maximum or minimum values comprise the only information needed for a particular problem.

In higher mathematics it can be shown that the X coordinate, or abscissa, of the maximum or minimum value is

469

$$x = \frac{-b}{2a}$$

In other words, if we divide minus the coefficient of the x term by twice the coefficient of the x^2 term, we have the X coordinate of the maximum or minimum point. If we substitute this value for x in the original equation, the result is the Y value or ordinate, which corresponds to the X value.

For example, we know that the graph of the equation

$$x^2 + 2x - 8 = y$$

passes through a minimum value because a is positive. To find the coordinates of the point where the parabola has its minimum value, we note that $a = 1$, $b = 2$, $c = -8$. From the rule given above, the X value of the minimum point is

$$x = \frac{-b}{2a}$$

$$x = \frac{-(2)}{2(1)}$$

$$x = -1$$

Substituting this value for x in the original equation, we have the value of the Y coordinate of the minimum point. Thus,

$$(-1)^2 + 2(-1) - 8 = y$$

$$1 - 2 - 8 = y$$

$$-9 = y$$

The minimum point is $(-1, -9)$. From the graph

in figure 16-1 (A), we see that these coordinates are correct. Thus, we can quickly and easily find the coordinates of the minimum or maximum point for any quadratic of the form $ax^2 + bx + c = 0$.

Practice problems. Without graphing, find the coordinates of the maximum or minimum points for the following equations and state whether they are maximum or minimum.

1. $2x^2 - 5x + 2 = 0$
2. $68 - 3x - x^2 = 0$
3. $3 + 7x - 6x^2 = 0$
4. $24x^2 - 14x = 3$

Answers:

1. $x = \dfrac{5}{4}$ Minimum

    $y = -\dfrac{9}{8}$

2. $x = -\dfrac{3}{2}$ Maximum

    $y = \dfrac{281}{4}$

3. $x = \dfrac{7}{12}$ Maximum

    $y = \dfrac{121}{24}$

4. $x = \dfrac{7}{24}$ Minimum

$y = -\dfrac{121}{24}$

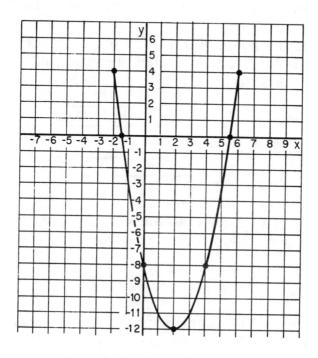

Figure 16-2. – Graph of $x^2 - 4x - 8 = 0$.

THE DISCRIMINANT

The roots of a quadratic equation may be classified in accordance with the following criteria:

1. Real or imaginary.

2. Rational or irrational.
3. Equal or unequal.

The task of discriminating among these possible characteristics to find the nature of the roots is best accomplished with the aid of the quadratic formula. The part of the quadratic formula which is used is called the DISCRIMINANT.

If the roots of a quadratic are denoted by the symbols r_1 and r_2, then the following relations may be stated:

$$r_1 = \frac{-b + \sqrt{b^2 - 4ac}}{2a}$$

$$r_2 = \frac{-b - \sqrt{b^2 - 4ac}}{2a}$$

We can show that the character of the roots is dependent upon the form taken by the expression

$$b^2 - 4ac$$

which is the quantity under the radical in the formula. This expression is the DISCRIMINANT of a quadratic equation.

IMAGINARY ROOTS

Since there is a radical in each root, there is a possibility that the roots could be imaginary. They are imaginary when the number under the radical in the quadratic formula is negative (less than 0). In other words, when the value of the discriminant is less than 0, the roots are imaginary.

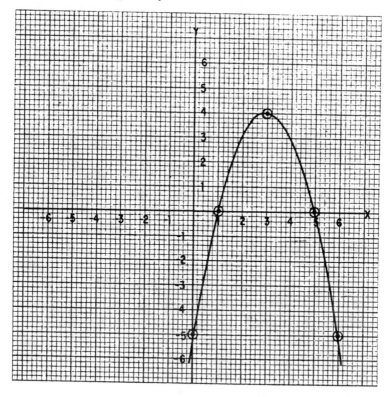

Figure 16-3.—Graph of 6x - 5 - x^2 = 0.

EXAMPLE:

$$x^2 + x + 1 = 0$$
$$a = 1, b = 1, c = 1$$
$$b^2 - 4ac = (1)^2 - 4(1)(1)$$
$$= 1 - 4$$
$$= -3$$

Thus, without further work, we know that the roots are imaginary.

CHECK: The roots are

$$r_1 = \frac{-1 + \sqrt{-3}}{2} \qquad r_2 = \frac{-1 - \sqrt{-3}}{2}$$

$$r_1 = -\frac{1}{2} + \frac{i\sqrt{3}}{2} \qquad r_2 = -\frac{1}{2} - \frac{i\sqrt{3}}{2}$$

We recognize both of these numbers as being imaginary.

We may also conclude that when one root is imaginary the other will also be imaginary. This is because the pairs of imaginary roots are always conjugate complex numbers. If one root is of the form a + ib, then a - ib is also a root. Knowing that imaginary roots always occur in pairs, we can conclude that a quadratic equation always has either two imaginary roots or two real roots.

Practice problems. Using the discriminant, state whether the roots of the following equations are real or imaginary:

1. $x^2 - 6x - 16 = 0$

2. $x^2 - 6x = -12$

3. $3x^2 - 10x + 50 = 0$

4. $6x^2 + x = 1$

Answers:

1. Real

2. Imaginary

3. Imaginary

4. Real

EQUAL OR DOUBLE ROOTS

If the discriminant $b^2 - 4ac$ equals zero, the radical in the quadratic formula becomes zero. In this case the roots are equal; such roots are sometimes called double roots.

Consider the equation

$$9x^2 + 12x + 4 = 0$$

Comparing with the general quadratic, we notice that

$$a = 9, \ b = 12, \text{ and } c = 4$$

The discriminant is

$$
\begin{aligned}
b^2 - 4ac &= 12^2 - 4(9)(4) \\
&= 144 - 144 \\
&= 0
\end{aligned}
$$

Therefore, the roots are equal.

CHECK: From the formula

$$r_1 = \frac{-12 + 0}{2(9)} \qquad r_2 = \frac{-12 - 0}{2(9)}$$

$$r_1 = -\frac{2}{3} \qquad r_2 = -\frac{2}{3}$$

The equality of the roots is thus verified.

The roots can be equal only if the trinomial is a perfect square. Its factors are equal. Factoring the trinomial in

$$9x^2 + 12x + 4 = 0$$

we see that

$$(3x + 2)^2 = 0$$

Since the factor $3x + 2$ is squared, we actually have

$$3x + 2 = 0$$

twice, and we have

$$x = -\frac{2}{3}$$

twice.

The fact that the same root must be counted twice explains the use of the term "double root." A double root of a quadratic equation is always rational because a double root can occur only when the radical vanishes.

REAL AND UNEQUAL ROOTS

When the discriminant is positive, the roots must be real. Also they must be unequal since equal roots occur only when the discriminant is zero.

Rational Roots

If the discriminant is a perfect square, the roots are rational. For example, consider the equation

$$3x^2 - x - 2 = 0$$

in which

$$2 = 3, \ b = -1, \ \text{and} \ c = -2$$

The discriminant is

$$b^2 - 4ac = (-1)^2 - 4(3)(-2)$$
$$= 1 + 24$$
$$= 25$$

We see that the discriminant, 25, is a perfect square. The perfect square indicates that the radical in the quadratic formula can be removed, that the roots of the equation are rational, and that the trinomial can be factored. In other words, when we evaluate the discriminant and find it to be a perfect square, we know that the trinomial can be factored.

Thus,

$$3x^2 - x - 2 = 0$$
$$(3x + 2)(x - 1) = 0$$

from which

$$3x + 2 = 0 \qquad x - 1 = 0$$
$$x = -\frac{2}{3} \qquad x = 1$$

We see that the information derived from the

discriminant is correct. The roots are real, unequal, and rational.

Irrational Roots

If the discriminant is not a perfect square, the radical cannot be removed and the roots are irrational.

Consider the equation

$$2x^2 - 4x + 1 = 0$$

in which

$$a = 2, \; b = -4, \text{ and } c = 1.$$

The discriminant is
$$b^2 - 4ac = (-4)^2 - 4(2)(1)$$
$$= 16 - 8$$
$$= 8$$

This discriminant is positive and not a perfect square. Thus the roots are real, unequal, and irrational.

To check the correctness of this information, we derive the roots by means of the formula. Thus,

$$x = \frac{-b \pm \sqrt{b^2 - 4ac}}{2a}$$

$$= \frac{4 \pm \sqrt{8}}{4}$$

$$= \frac{2 \pm \sqrt{2}}{2}$$

$$x = 1 + \frac{\sqrt{2}}{2} \text{ or } x = 1 - \frac{\sqrt{2}}{2}$$

This verifies the conclusions reached in evaluating the discriminant. When the discriminant is a positive number, not a perfect square, it is useless to attempt to factor the trinomial. The formula is needed to find the roots. They will be real, unequal, and irrational.

SUMMARY

The foregoing information concerning the discriminant may be summed up in the following four rules:

1. If $b^2 - 4ac$ is a perfect square or zero, the roots are rational; otherwise they are irrational.

2. If $b^2 - 4ac$ is negative (less than zero), the roots are imaginary.

3. If $b^2 - 4ac$ is zero, the roots are real, equal, and rational.

4. If $b^2 - 4ac$ is greater than zero, the roots are real and unequal.

Practice problems. Determine the character of the roots of each of the following equations:

1. $x^2 - 7x + 12 = 0$

2. $9x^2 - 6x + 1 = 0$

3. $2x^2 - x + 1 = 0$

4. $2x - 2x^2 + 6 = 0$

Answers:

1. Real, unequal, rational

2. Real, equal, rational

3. Imaginary

4. Real, unequal, irrational

GRAPHICAL INTERPRETATION OF ROOTS

When a quadratic is set equal to y and the resulting equation is graphed, the graph will reveal the character of the roots, but it may not reveal whether the roots are rational or irrational.

Consider the following equations:

1. $x^2 + 6x - 3 = y$

2. $x^2 + 6x + 9 = y$

3. $x^2 + 6x + 13 = y$

The graphs representing these equations are shown in figure 16-4.

We recall that the roots of the equation are the values of x at those points where y is zero. Y is zero on the graph anywhere along the X axis. Thus, the roots of the equation are the positions where the graph crosses the X axis. In parabola No. 1 (fig. 16-4) we see immediately that there are two roots to the equation and that they are unequal. These roots appear to be -6.5 and 0.5. Algebraically, we find them to be the irrational numbers

$$-3 + 2 \sqrt{3} \text{ and } -3 - 2 \sqrt{3}.$$

For equation No. 2 (fig. 16-4), the parabola

481

just touches the X axis at x = -3. This means that both roots of the equation are the same— that is, the root is a double root. At the point where the parabola touches the X axis, the two roots of the quadratic equation have moved

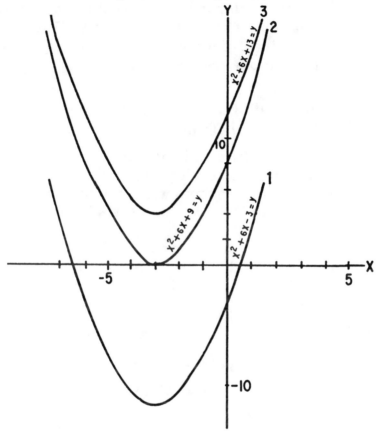

Figure 16-4.—Graphical interpretation of roots.

together and the two points of intersection of the parabola and the X axis are coincident. The quantity -3 as a double root agrees with the algebraic solution.

When the equation No. 3 (fig. 16-4) is solved algebraically, we see that the roots are -3 + 2i and -3 - 2i. Thus they are imaginary. Parabola No. 3 does not cross the X axis. When this situation occurs, imaginary roots are implied. Only equations having real roots will have graphs that cross or touch the X axis. Thus we may determine from the graph of an equation whether the roots are real or imaginary.

VERBAL PROBLEMS INVOLVING QUADRATIC EQUATIONS

Many practical problems give rise to quadratic equations. In such problems it often happens that one of the roots will have no meaning. We must select the root that satisfies the conditions of the problem.

Consider the following example: The length of a plot of ground exceeds its width by 7 ft and the area of the plot is 120 sq ft. What are the dimensions?

SOLUTION:

$$Let\ x\ =\ length$$
$$y\ =\ width$$

then

$$x - y = 7 \qquad (1)$$

and

$$xy = 120 \qquad (2)$$

Solving (1) for y, y = x - 7

Substituting (x - 7 for y in (2)

$$x(x - 7) = 120$$

Therefore

$$x^2 - 7x - 120 = 0$$

$$(x - 15)(x + 8) = 0$$

$$x = 15, \qquad x = -8$$

Thus, length = +15 or -8.

But the length obviously cannot be a negative value. Therefore, we reject -8 as a value for x and use only the positive value, +15. Then from equation (1),

$$15 - y = 7$$

$$y = 8$$

Length = 15, Width = 8

Practice problems. Solve the following problems by forming quadratic equations:

1. A rectangular plot is 8 yd by 24 yd. If the length and width are increased by the same amount, the area is increased by 144 sq yd. How much is each dimension increased?

2. Two cars travel at uniform rates of speed

over the same route a distance of 180 mi. One goes 5 mph slower than the other and takes 1/2 hr longer to make the run. How fast does each car travel?

Answers:

1. Length and width are each increased by 4 yd.

2. Faster car: 45 mph.
   Slower car: 40 mph.

CHAPTER 17

PLANE FIGURES

The discussion of lines and planes in chapter 1 of this course was limited to their consideration as examples of sets. The present chapter is concerned with lines, angles, and areas as found in various plane (flat) geometric figures.

LINES

In the strictly mathematical sense, the term "line segment" should be used whenever we refer to the straight line joining some point A to some other point B. However, since the straight lines comprising geometric figures have clearly designated end points, we may simplify our terminology. Throughout the remaining chapters of this course, the general term "line" is used to designate straight line segments, unless stated otherwise.

TYPES OF LINES

The two basic types of lines in geometry are straight lines and curved lines. A curved line joining points A and B is designated as "curve

AB." (See fig. 17-1.) If curve AB is an arc of a circle, it may be designated as "arc AB."

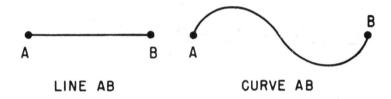

LINE AB   CURVE AB

Figure 17-1.—Straight and curved lines.

The term "broken line" in mathematics means a series of two or more straight segments connected end-to-end but not running in the same direction. In mathematics, a series of short, straight segments with breaks between them, which would form a single straight line if joined end-to-end, is a DASHED LINE. (See fig. 17-2.)

BROKEN LINE   DASHED LINE

Figure 17-2.—Broken and dashed lines.

ORIENTATION

Straight lines may be classified in terms of their orientation to the observer's horizon or in terms of their orientation to each other. For example, lines in the same plane which run beside each other without meeting at any point, no matter how far they are extended, are PARALLEL. (See fig. 17-3 (A).) Lines in the same plane which are not parallel are OBLIQUE. Oblique lines meet to form angles (discussed in the following section). If two oblique lines cross or meet in such a way as to form four equal angles, as in figure 17-3 (B), the lines are PERPENDICULAR. This definition includes the case in which only one angle is formed, such as angle AEC in figure 17-3 (C). By extending line AE to form line AD, and extending CE to form CB, four equal angles (AEC, CED, DEB, and BEA) are formed.

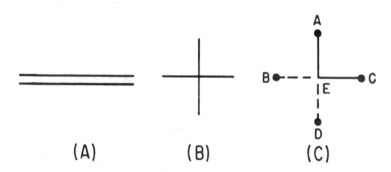

(A)　　　　　(B)　　　　　(C)

Figure 17-3.—(A) Parallel lines; (B) and (C) perpendicular lines.

Lines parallel to the horizon are HORIZONTAL. Lines perpendicular to the horizon are VERTICAL.

ANGLES

Lines which meet or cross each other are said to INTERSECT. Angles are formed when two straight lines intersect. The two lines which form an angle are its SIDES, and the point where the sides intersect is the VERTEX. In figure 17-4, the sides of the angles are AV and BV, and the vertex is V in each case. Figure 17-4 (A) is an ACUTE angle; (B) is an OBTUSE angle.

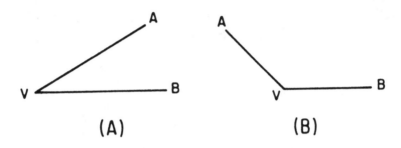

(A)   (B)

Figure 17-4.—(A) Acute angle; (B) obtuse angle.

CLASSIFICATION BY SIZE

When the sides of an angle are perpendicular to each other, the angle is a RIGHT angle. This term is related to the Latin word "rectus," which may be translated "erect" or "upright." Thus, if one side of a right angle is horizontal, the other side is erect or upright.

The size of an angle refers to the amount of separation between its sides, and the unit of angular size is the angular DEGREE. A right angle contains 90 degrees, abbreviated 90°. An angle smaller than a right angle is acute; an angle larger than a right angle is obtuse. Therefore, acute angles are angles of less than 90°, and obtuse angles are angles between 90° and

180°.

If side AV in figure 17-5 (A) is moved downward, the size of the obtuse angle AVB is increased. If side AV is moved so far that it coincides with (lies on top of) CV as in figure 17-5 (B), an angle is formed which is equal to the sum of two right angles. The special angle thus formed (AVB) is a straight angle, so called because it is visually indistinguishable from a straight line.

GEOMETRIC RELATIONSHIPS

Angles are often classified by their relationship to other angles or to other parts of a geometric figure. For example, angles 1 and 3 in

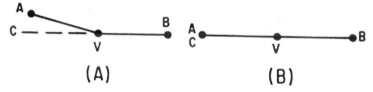

Figure 17-5.—(A) Large obtuse angle;
(B) straight angle.

figure 17-6 are VERTICAL angles, so called because they share a common vertex. Angles 2 and 4 are also vertical angles. Lines which cross, as in figure 17-6, always form two pairs of vertical angles, and the vertical angles thus formed are equal in pairs; that is, angle 1 equals angle 3, and angle 2 equals angle 4.

Angles 1 and 2 in figure 17-6 are ADJACENT angles. Other pairs of adjacent angles in figure 17-6 are 2 and 3, 3 and 4, and 1 and 4. In

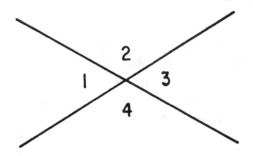

Figure 17-6.—Vertical angles.

the sense used here, adjacent means side by side, not merely close together or touching. For example, angles 1 and 3 are not adjacent angles even though they touch each other.

COMPLEMENTS AND SUPPLEMENTS

Two angles whose sum is 90° are complementary. For example, a 60° angle is the complement of a 30° angle, and conversely. "Conversely" is a mathematical word meaning "vice versa." Two angles whose sum is 180° are supplementary. For example, a 100° angle is the supplement of an 80° angle, and conversely.

Practice problems.

1. Describe the angle which is the complement of an acute angle.

2. Describe the angle which is the supplement of a right angle.

3. If two equal angles are complementary, each contains how many degrees?

491

4. Find the size of an angle which is twice as large as its own complement.

(Hint: If x is the angle, then 90° - x is its complement.)

Answers:

1. Acute

2. Right

3. 45°

4. 60°

GEOMETRIC FIGURES

The discussion of geometric figures in this chapter is limited to polygons and circles. A POLYGON is a plane closed figure, the sides of which are all straight lines. Among the polygons discussed are triangles, parallelograms, and trapezoids.

TRIANGLES

A triangle is a polygon which has three sides and three angles. In general, any polygon has as many angles as it has sides, and conversely.

Parts of a Triangle

Each of the three angles of a triangle is a VERTEX; therefore, every triangle has three vertices. The three straight lines joining the vertices are the SIDES (sometimes called legs), and the side upon which the triangle rests is its BASE, often designated by the letter b. This definition assumes that the standard position of a triangle drawn for general discussion is as shown in figure 17-7, in which the triangle is lying on one of its sides. The vertex opposite the base is the highest point of a triangle in standard position, and is thus called the APEX.

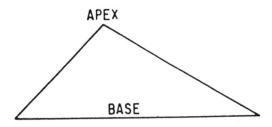

Figure 17-7.—Triangle in standard position.

A straight line perpendicular to the base of a triangle, joining the base to the apex, is the ALTITUDE, often designated by the letter a. The altitude is sometimes referred to as the height, and is then designated by the letter h. Figure 17-8 (B) shows that the apex may not be situated directly above the base. In this case, the base must be extended, as shown by the dashed line, in order to drop a perpendicular from the apex to the base. Mathematicians often use the term "drop a perpendicular." The meaning is the same as "draw a straight, perpendicular line."

In general, the geometrical term "distance from a point to a line" means the length of a perpendicular dropped from the point to the line. Many straight lines could be drawn from a line to a point not on the line, but the shortest of these is the one we use in measuring the

493

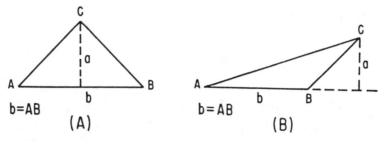

Figure 17-8.—(A) Interior altitude line; (B) exterior altitude line.

distance from the point to the line. The short-est one is perpendicular to the line.

Perimeter and Area

The PERIMETER of a triangle is the sum of the lengths of its sides. In less precise terms, this is sometimes stated as "the distance around the triangle." If the three sides are labeled a, b, and c, the perimeter P can be found by the following formula:

$$P = a + b + c$$

The area of a triangle is the space bounded (enclosed) by its sides. The formula for the area can be found by using a triangle which is part of a rectangle. In figure 17-9, triangle ABC is one-half of the rectangle. Since the area of the rectangle is a times b (that is, ab), the area of the triangle is given by the follow-ing formula:

$$\text{Area} = \frac{1}{2} \text{ ab}$$

Written in terms of h, representing height,

the formula is:

$$A = \frac{1}{2} bh$$

This formula is valid for every triangle, including those with no two sides perpendicular.

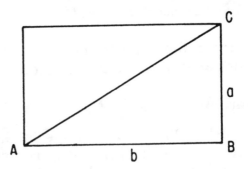

Figure 17-9.—Area of a triangle.

Practice problems. Find the perimeter and area of each of the triangles in figure 17-10.

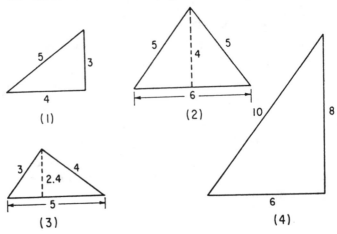

Figure 17-10.—Perimeters and areas of triangles.

495

Answers:

1. P = 12 units
   A = 6 square units

2. P = 16 units
   A = 12 square units

3. P = 12 units
   A = 6 square units

4. P = 24 units
   A = 24 square units

CAUTION: The concept of area is meaningless if the units of the multiplied dimensions are not the same. For example, if the base of a triangle is 2 feet long and the altitude is 6 inches long, the area might be carelessly stated as $\frac{1}{2}$ (6) (2). However, the units must be considered in order to decide whether the answer is in square feet or square inches. When the units are considered, we realize that the correct answer is

$$\frac{1}{2} \text{ (6 in.) (24 in.)} = 72 \text{ sq in.}$$

$$\frac{1}{2} \left(\frac{1}{2} \text{ ft}\right) \text{ (2 ft)} = \frac{1}{2} \text{ sq ft}$$

Special Triangles

The classification of triangles depends upon their special characteristics, if any. For example, a triangle may have all three of its sides equal in length; it may have two equal sides and a third side which is longer or shorter than the other two; it may contain a right angle or an obtuse angle. If it has none of these special characteristics, it is a SCALENE triangle. A scalene triangle has no two of its sides equal and no two of its angles equal.

RIGHT TRIANGLE.—If one of the angles of a triangle is a right angle, the figure is a right triangle. The sides which form the right angle are the LEGS of the triangle, and the third side (opposite the right angle) is the HYPOTENUSE.

The area of a right triangle is always easy to determine. If the base of the triangle is one of its legs, as in figure 17-10 (4), the other leg is the altitude. If the hypotenuse is acting as the base, as in figure 17-10 (3), the triangle can be turned until one of its legs is the base, as in figure 17-10 (1). If the triangle is not known to be a right triangle, then the altitude must be given, as in figure 17-10 (2), in order to calculate the area.

Any triangle whose sides are in the ratio of 3:4:5 is a right triangle. Thus, triangles with sides as follows are right triangles:

Side 1	Side 2	Side 3
3	4	5
6	8	10
12	16	20
3x	4x	5x

(x is any positive number)

In addition to the 3-4-5 triangle, two other types of right triangles occur frequently. Any triangle having one 30° angle and one 60° angle is a right triangle; that is, its third angle is 90°. Any triangle having two 45° angles is a right triangle.

ISOSCELES TRIANGLE.—A triangle having two of its sides equal in length is an ISOSCELES triangle. Since the length of the side opposite an angle is determined by the size of the angle, the isosceles triangle has two equal angles. In figure 17-11 (A), triangle ABC is isosceles. Sides AC and BC are equal in length, and angles A and B are equal.

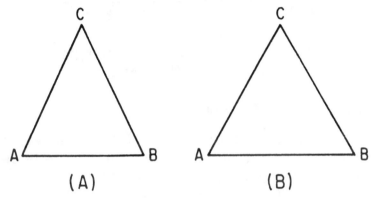

Figure 17-11.—(A) Isoceles triangle; (B) equilateral triangle.

Figure 17-11 (B) illustrates an EQUILATERAL triangle, which is a special case of an isosceles triangle. An equilateral triangle has all three of its sides equal in length. Since the lengths of the sides are directly related to the size of the angles opposite them, an equilateral triangle is also equiangular; that is, all three of its angles are equal.

OBLIQUE TRIANGLES.—Any triangle containing no right angle is an OBLIQUE triangle. Figure 17-12 illustrates two possible configurations, both of which are oblique triangles. An oblique triangle which contains an obtuse angle is often called an OBTUSE triangle.

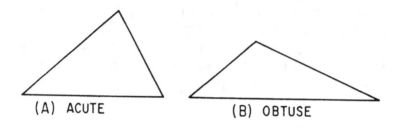

Figure 17-12.—Oblique triangles.
(A) Acute; (B) obtuse.

Sum of the Angles

The sum of the angles in any triangle is 180°. For example, if one of the angles is 40° and another is 20°, the third angle is 120°. It is this relationship that justifies the statements made in the preceding section concerning 45° triangles and 30°-60°-90° triangles. If two of the angles are 45° each, then the third angle is 180° - (45°+45°) and the figure is a right triangle. If one angle is 60° and another is 30°, the third angle is 90° and the figure is a right triangle.

QUADRILATERALS

A QUADRILATERAL is a polygon with four sides. The parts of a quadrilateral are its sides, its four angles, and its two DIAGONALS. A diagonal is a straight line joining two alternate vertices of a polygon. Figure 17-13 illustrates the parts of a quadrilateral, in which AC and DB are the diagonals.

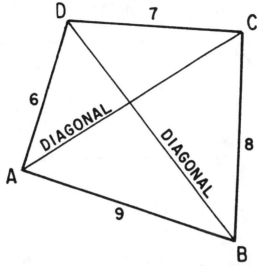

Figure 17-13.—Parts of a quadrilateral.

Perimeter and Area

The perimeter of a quadrilateral is the sum of the lengths of its sides. For example, the perimeter of the quadrilateral in figure 17-13 is 30 units.

The area of a quadrilateral can be found by dividing it into triangles and summing the areas of the triangles. However, the altitudes of the triangles are usually difficult to calculate unless the quadrilateral has at least one pair of parallel sides.

Parallelograms

A PARALLELOGRAM is a quadrilateral in which the opposite sides are parallel. For example, in the parallelogram in figure 17-14, side AB is parallel to side CD. Furthermore, side BC is parallel to side AD.

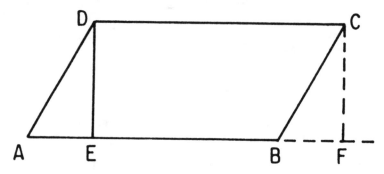

Figure 17-14.—A parallelogram.

Since lines AB and CD are parallel, lines DE and CF (both perpendicular to line AF in figure 17-14) are equal. Angles DAE and CBF in figure 17-14 are equal, because a straight line cutting two parallel lines, such as AD and BC, forms equal angles with the parallel lines. Thus, triangles AED and BFC are equal, and line AD equals line BC. Therefore we have proved that the opposite sides of a parallelogram are equal. If all four of the sides are the same length, the parallelogram is a RHOMBUS.

In addition to the equality of the opposite sides, the opposite angles of a parallelogram are also equal. For example, angle DAB equals angle BCD in figure 17-14, and angle ADC equals angle ABC.

RECTANGLES AND SQUARES.—When all of the angles of a parallelogram are right angles, it is a RECTANGLE. A rectangle with all four of its sides the same length is a SQUARE. Thus a square is a rhombus having 90° angles. Every square is a rectangle, and every rectangle is a parallelogram. Notice that the reverse of this statement is not true.

The area of a rectangle is found by multiplying its length times its width. Therefore, if each side of a square has length s, the area of the square is s^2.

Written as formulas, these areas are as follows:

Rectangle: $A = lw$

or $\qquad A = bh$, where b = base,

$\qquad\qquad\qquad h$ = height

Square: $\quad A = s^2$

AREA.—The area of a parallelogram can be found by dividing it into rectangles and triangles. For example, in figure 17-14 the area of the parallelogram is the sum of the areas of triangle AED and figure EBCD. Since triangle AED is equal to triangle BFC, the sum of AED and EBCD is equal to the sum of BFC and EBCD. Thus the area of parallelogram ABCD is the same as the area of rectangle EFCD. Since the area of EFCD is DC multiplied by DE, and DC has the same length as AB, we conclude that the area of a parallelogram is the product of its base times its altitude. Written as a formula, this is

$$A = ba$$

or

$$A = bh, \text{ where } h \text{ is height}$$

Trapezoids

A TRAPEZOID is a quadrilateral in which two sides are parallel and the other two sides are not parallel. By orienting a trapezoid so that its parallel sides are horizontal, we may call the parallel sides bases. Observe that the bases of a trapezoid are not equal in length. (See fig. 17-15.)

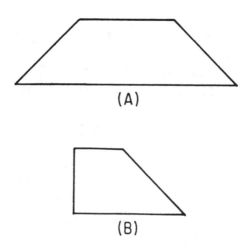

(A)

(B)

Figure 17-15.—Typical trapezoids.

The area of a trapezoid may be found by separating it into two triangles and a rectangle, as in figure 17-16. The total area A of the trapezoid is the sum of A_1 plus A_2 plus A_3, and is calculated as follows:

$$A = A_1 + A_2 + A_3$$

$$= \frac{1}{2}ha + hb_1 + \frac{1}{2}hc$$

$$= \frac{1}{2}h(a + 2b_1 + c)$$

$$= \frac{1}{2}h(a + b_1 + c) + b_1$$

$$= \frac{1}{2}h(b + b_1)$$

Thus the area of a trapezoid is equal to one-half the altitude times the sum of the bases.

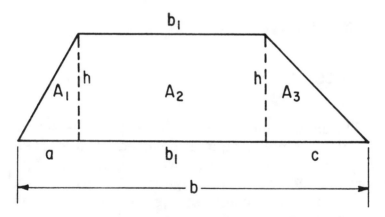

Figure 17-16.—Area of a trapezoid.

Practice problems. Find the area of each of the following figures:

1. Rhombus; base 4 in., altitude 3 in.

2. Rectangle; base 6 ft, altitude 4 ft

3. Parallelogram; base 10 yd, altitude 12 ft

4. Trapezoid; bases 6 ft and 4 ft, altitude 2 yd.

Answers:

1. 12 sq in.

2. 24 sq ft

3. 40 sq yd

4. 30 sq ft

CIRCLES

The mathematical definition of a circle states that it is a plane figure bounded by a curved line, every point of which is equally distant from the center of the figure. The parts of a circle are its circumference, its radius, and its diameter.

Parts of a Circle

The CIRCUMFERENCE of a circle is the line that forms its outer boundary. Circumference is the special term used in referring to the "perimeter" of a circle. (See fig. 17-17.) A RADIUS of a circle is a line joining the center to a point on the circumference, as shown in figure 17-17. A straight line joining two points on the circumference of a circle, and passing through the center, is a DIAMETER. A straight line which touches the circle at just one point is a TANGENT. A tangent is perpendicular to a radius at the point of tangency.

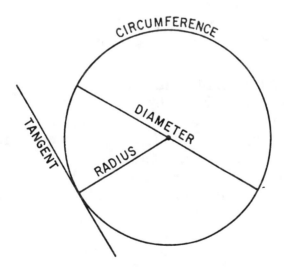

Figure 17-17.—Parts of a circle.

An ARC is a portion of the circumference of a circle. A CHORD is a straight line joining the end points of any arc. The portion of the area of a circle cut off by a chord is a SEGMENT of the circle, and the portion of the circle's area cut off by two radii (radius lines) is a SECTOR. (See fig. 17-18.)

Formulas for Circumference and Area

The formula for the circumference of a circle is based on the relationship between the circumference and the diameter. This comparison can be made experimentally by marking the edge of a circular object, such as a coin, and rolling it (without slippage) along a flat surface. (See fig. 17-19.)

The distance from the initial position to the final position of the disk in figure 17-19 is approximately 3.14 times as long as the diameter of the disk. With any circle, this is always found to be the case; but it is not possible to give the value of C/d (circumference divided by diameter) exactly. The ratio C/d is represented by the symbol π, which is the Greek letter pi. Thus we have the following equations:

$$\frac{C}{d} = \pi$$

$$C = \pi d$$

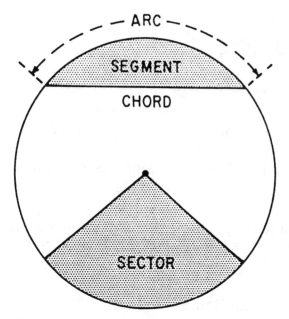

Figure 17-18.—Arc, chord, segment, and sector.

This formula states that the circumference of a circle is π times the diameter. Notice that it could be written as

$$C = 2r \cdot \pi \text{ or } C = 2\pi r$$

since the diameter d is the same as 2r (twice the radius).

Although the value of π is not exactly equal to any of the numerical expressions which are sometimes used for it, the ratio is very close to 3.14. If extreme accuracy is required, 3.1416 is used as an approximate value of π. Many calculations involving π are satisfactory if the fraction 22/7 is used as the value of π.

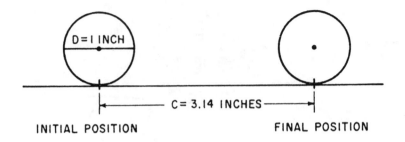

Figure 17-19.—Measuring the circumference
of a circle.

Practice problems. Calculate the circumference of each of the following circles, using 22/7 as the value of π:

1. Radius = 21 in.

2. Diameter = 7.28 in.

3. Radius = 14 ft

4. Diameter = 2.8 yd

Answers:

1. 132 in.

2. 22.88 in.

3. 88 ft

4. 8.8 yd

AREA.—The area of a circle is found by multiplying the square of its radius by π. The formula is written as follows:

$$A = \pi r^2$$

EXAMPLE: Find the area of a circle whose diameter is 4 ft.

SOLUTION: The radius is one-half the diameter. Therefore,

$$r = \frac{1}{2}(4 \text{ ft})$$

$$= 2 \text{ ft}$$

$$A = \pi r^2 = \pi(2 \text{ ft})^2$$

$$A = 3.14 \ (4 \text{ sq ft})$$

$$= 12.56 \text{ sq ft}$$

Practice problems. Find the area of each of the following circles:

1. Radius = 7 in.
2. Diameter = 42 mi
3. Diameter = 2.8 ft
4. Radius = 14 yd

Answers:

1. A = 154 sq in.
2. A = 1,386 sq mi
3. 6.16 sq ft
4. 616 sq yd

Concentric Circles

Circles which have a common center are said to be CONCENTRIC. (See fig. 17-20.)

The area of the ring between the concentric circles in figure 17-20 is calculated as follows:

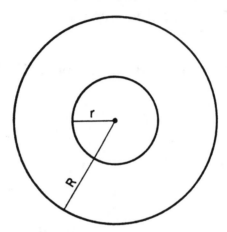

Figure 17-20.—Concentric circles.

Let R = radius of large circle

r = radius of small circle

A_R = area of large circle

A_r = area of small circle

A = area of ring

Then $A = A_R - A_r$

$= \pi R^2 - \pi r^2$

$= \pi(R^2 - r^2)$

511

Notice that the last expression is the difference of two squares. Factoring, we have

$$A = \pi(R + r)(R - r)$$

Therefore, the area of a ring between two circles is found by multiplying π times the product of the sum and difference of their radii.

Practice problems. Find the areas of the rings between the following concentric circles:

1. R = 4 in.   2. R = 6 ft

   r = 3 in.   r = 2 ft

   Answers:

1. 22 sq in.   2. 100.6 sq ft

CHAPTER 18

GEOMETRIC CONSTRUCTIONS AND SOLID FIGURES

CONSTRUCTIONS

From the standpoint of geometry, a CONSTRUCTION may involve either the process of building up a figure or that of breaking down a figure into smaller parts. Some typical constructions are listed as follows:

1. Dividing a line into equal segments.
2. Erecting the perpendicular bisector of a line.
3. Erecting a perpendicular at any point on a line.
4. Bisecting an angle.
5. Constructing an angle.
6. Finding the center of a circle.
7. Constructing an ellipse.

EQUAL DIVISIONS ON A LINE

A line may be divided into any desired number of equal segments by the method shown in figure 18-1.

Suppose that line AB (fig. 18-1) is to be divided into seven equal segments. Draw line AC at any convenient angle with AB and mark

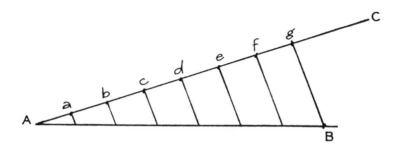

Figure 18-1.—Dividing a line into
equal segments.

off seven spaces of some convenient length, say
1/2 inch, on it.   Extend AC, if necessary, in
order to get seven intervals of the chosen length
on it.   This produces the points a, b, c, d, e, f,
and g, as shown in figure 18-1.   Draw a line
from g to B, and then draw lines parallel to gB,
starting at each of the points a, b, c, d, e, and f.
The segments of AB cut off by these lines are
equal in length.

It is frequently necessary to rule a pre-
determined number of lines on a blank sheet of
material.   This may be done by a method based
on the foregoing discussion. For example, suppose that the sheet of typing paper in figure
18-2 is to be divided into 24 equal spaces.

The 12-inch ruler is laid across the paper
at an angle, in such a way that the ends of the
ruler coincide with the top and bottom edges of
the paper.   There are 24 spaces, each 1/2 inch
wide, on a 12-inch ruler.   Therefore, we mark
the paper beside each 1/2-inch division marker
on the ruler.   After removing the ruler, we
draw a line through each of the marks on the
paper, parallel to the top and bottom edges of
the paper.

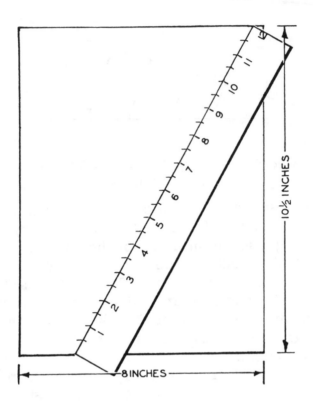

Figure 18-2.—Ruling equal spaces on
a sheet of paper.

PERPENDICULAR BISECTOR OF A LINE

To bisect a line or an angle means to divide
it into two equal parts. A line may be bisected
satisfactorily by measurement, or by a geometric method. If the measuring instrument
does not reach the full length of the line, proceed as follows:

1. Starting at one end, measure about half
the length of the line and make a mark.

2. Starting at the other end, measure exactly

515

the same distance as before and make a second mark.

3. The bisector of the line lies halfway between these two marks.

The geometric method of bisecting a line is not dependent on measurement. It is based upon the fact that all points equally distant from the ends of a straight line lie on the perpendicular bisector of the line.

Bisecting a line geometrically requires the use of a mathematical compass, which is an instrument for drawing circles and comparing distances. If a line AB is to be bisected as in figure 18-3, the compass is opened until the distance between its points is more than half as long as AB. Then a short arc is drawn above the approximate center of the line and another below, using A as the center of the arcs' circle. (See fig. 18-3.)

Two more short arcs are drawn, one above and one below the approximate center of line AB, this time using B as the center of the arcs' circle.

The two arcs above line AB are extended until they intersect, forming point C, and the two arcs below line AB intersect to form point D. The line joining point C and point D is the perpendicular bisector of line AB.

PERPENDICULAR AT ANY POINT ON A LINE

Figure 18-4 shows a line AB with point C between A and B. A perpendicular to AB is erected at C as follows:

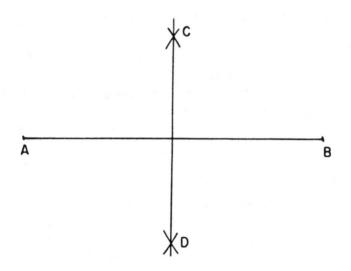

Figure 18-3.—Bisecting a line geometrically.

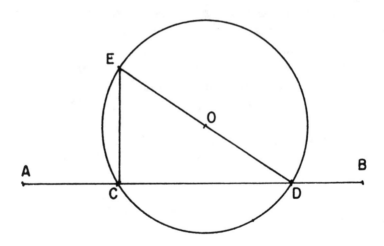

Figure 18-4.—Erecting a perpendicular
at a point.

1. Using any convenient point above the line (such as O) as a center, draw a circle with ra-

dius OC. This circle cuts AB at C and at D.

2. Draw line DO and extend it to intersect the circle at E.

3. Draw line EC. This line is perpendicular to AB at C.

BISECTING AN ANGLE

Let angle AOB in figure 18-5 be an angle which is to be bisected. Using O as a center and any convenient radius, draw an arc intersecting OA and a second arc intersecting OB. Label these intersections C and D.

Using C and D as centers, and any convenient radius, draw two arcs intersecting halfway between lines OA and OB. A line from O through the intersection of these two arcs is the bisector of angle AOB.

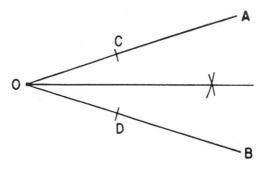

Figure 18-5.—Bisecting an angle.

SPECIAL ANGLES

Several special angles may be constructed by geometric methods, so that an instrument for measuring angles is not necessary in these special cases.

518

Figure 18-4 illustrates a method of constructing a right angle, DCE, by inscribing a right triangle in a semicircle. But an alternate method is needed for those situations in which drawing circles is inconvenient. The method described herein makes use of a right triangle having its sides in the ratio of 3 to 4 to 5. It is often used in laying out the foundations of buildings. The procedure is as follows:

1. A string is stretched as shown in figure 18-6, forming line AC. The length of AC is 3 feet.

2. A second string is stretched, crossing line AC at A, directly above the point intended as the corner of the foundation. Point D on this line is 4 feet from A.

3. Attach a third string, 5 feet long, at C and D. When AC and AD are spread so that line CD is taut, angle DAC is a right angle.

A 60° angle is constructed as shown in figure 18-7. With AB as a radius and A and B as

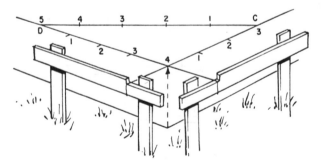

Figure 18-6.—Constructing a right angle by the 3-4-5 method.

centers, draw arcs intersecting at C. When A and B are connected to C by straight lines, all three angles of triangle ABC are 60° angles.

The special angles already discussed are used in constructing 45° and 30° angles. A 90° angle is bisected to form two 45° angles, and a 60° angle is bisected to form two 30° angles.

FINDING THE CENTER OF A CIRCLE

It is sometimes necessary to find the center of a circle of which only an arc or a segment is given. (See fig. 18-8.)

From any point on the arc, such as A, draw two chords intersecting the arc in any two points, such as B and C. With the points A, B, and C as centers, use any convenient radius and draw short intersecting arcs to form the perpendicular bisectors of chords AC and AB. Join the intersecting arcs on each side of AC, obtaining line MP, and join the arcs on each side of AB, obtaining line NQ. The intersection of MP and NQ is point O, the center of the circle.

ELLIPSES

An ellipse of specified length and width is constructed as follows:

1. Draw the major axis, AB, and the minor axis, CD, as shown in figure 18-9.

2. On a straightedge or ruler, mark a point (labeled a in the figure) and from this point measure one-half the length of the minor axis and make a second mark (b in figure 18-9). From point a, measure one-half the length of the major axis and make a third mark (c in the figure).

3. Place the straightedge on the axes so that

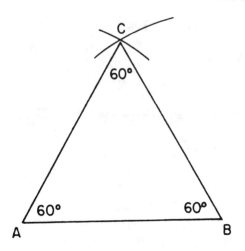

Figure 18-7.—Constructing 60° angles.

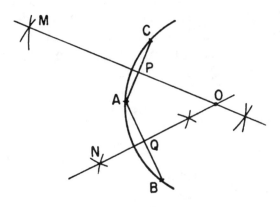

Figure 18-8.—Finding the center
of a circle.

b lies on the major axis and c lies on the minor
axis. Mark the paper with a dot beside point a.
Reposition the straightedge, keeping b on the
major axis and c on the minor axis, and make a
dot beside the new position of a.

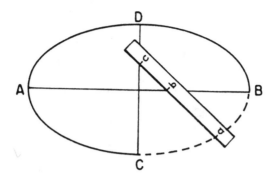

Figure 18-9.—Constructing an ellipse.

4. After locating enough dots to see the elliptical pattern, join the dots with a smooth curve.

SOLID FIGURES

The plane figures discussed in chapter 17 of this course are combined to form solid figures. For example, three rectangles and two triangles may be combined as shown in figure 18-10. The flat surfaces of the solid figure are its FACES; the top and bottom faces are the BASES, and the faces forming the sides are the LATERAL FACES.

Some solid figures do not have any flat faces, and some have a combination of curved surfaces and flat surfaces. Examples of solids with curved surfaces include cylinders, cones, and spheres.

PRISMS

The solid shown in figure 18-10 is a PRISM. A prism is a solid with three or more lateral faces which intersect in parallel lines.

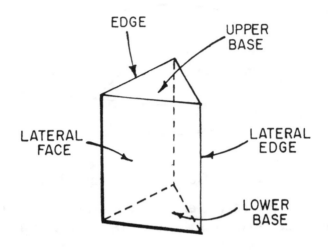

Figure 18-10.—Parts of a solid figure.

Types of Prisms

The name of a prism depends upon its base polygons. If the bases are triangles, as in figure 18-10, the figure is a TRIANGULAR prism. A RECTANGULAR prism has bases which are rectangles.

If the bases of a prism are perpendicular to the planes forming its lateral faces, the prism is a RIGHT prism.

A PARALLELEPIPED is a prism with parallelograms for bases. Since the bases are parallel to each other, this means that they cut the lateral faces to form parallelograms. Therefore, in a parallelepiped, all of the faces are parallelograms. If a parallelepiped is a right prism, and if its bases are rectangles, it is a rectangular solid. A CUBE is a rectangular solid in which all of the six rectangular faces are squares.

Parts of a Prism

The parts of a prism are shown in figure 18-10. The line formed by the joining of two faces of a prism is an EDGE. If the two faces forming an edge are lateral faces, the edge thus formed is a LATERAL EDGE.

Surface Area and Volume

The SURFACE AREA of a prism is the sum of the areas of all of its faces, including the bases. The VOLUME of a prism may be considered as the sum of the volumes of many thin wafers, each having a thickness of one unit and a shape that duplicates the shape of the base. (See fig. 18-11.)

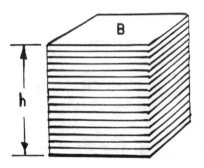

Figure 18-11.—Volume of a prism.

The wafers which comprise the prism in figure 18-11 all have the same area, which is the area of the base. Therefore, the volume of the prism is found by multiplying the area of the base times the number of wafers. Since each wafer is 1 inch thick, the number of wafers is the same as the height of the prism in inches. The resulting formula for the volume of a prism, using B to represent the area of the base and h to represent the height, is as follows:

$$V = Bh$$

When a prism has lateral edges which are not perpendicular to the bases, the height of the prism is the perpendicular distance between the bases. (See fig. 18-12.) The formula for the volume remains the same, even though the prism is no longer a right prism.

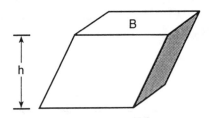

Figure 18-12.—Height of a prism
which is not a right prism.

CIRCULAR CYLINDERS

Any surface may be considered as the result of moving a straight line in a direction at right angles to its length. For example, suppose that the stick of charcoal in figure 18-13 is moved from position AB to position CD by dragging it across the paper. The broad mark made by the charcoal represents a plane surface. The surface is said to be "generated" by moving line AB.

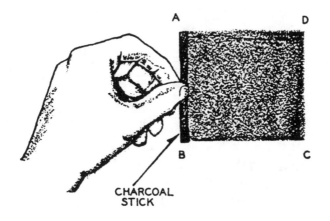

Figure 18-13.—Surface generated by
a moving line.

The movement of the line in figure 18-13 may be controlled by requiring that its lower end trace a particular path. For example, if line AB moves so as to trace an ellipse as in figure 18-14 (A), a cylindrical surface is generated by the line. This surface, shown in figure 18-14 (B), is an elliptical cylinder.

Any line in the surface, parallel to the generating line, such as CD or EF in figure 18-14 (B), is an ELEMENT of the cylinder. If the elements are perpendicular to the bases, the cylinder is a RIGHT CYLINDER. If the bases are circles, the cylinder is a CIRCULAR CYLINDER. Figure 18-14 (C) illustrates a right circular cylinder. Line O-O', joining the centers of the bases of a right circular cylinder, is the AXIS of the cylinder.

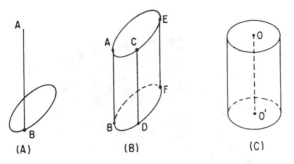

Figure 18-14.—(A) Line generating a cylinder;
(B) elliptical cylinder;
(C) circular cylinder.

Surface Area and Volume

The lateral area of a cylinder is the area of its curved surface, excluding the area of its bases. Figure 18-15 illustrates an experimental method of determining the lateral area of a right circular cylinder.

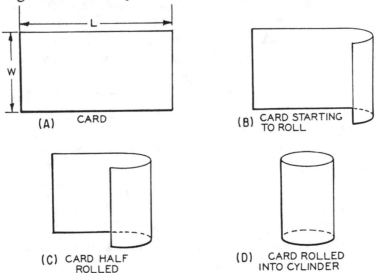

Figure 18-15.—Lateral area of a cylinder.

The card of length L and width W in figure 18-15 is rolled into a cylinder. The height of the cylinder is W and the circumference is L. The lateral area is the same as the original area of the card, LW. Therefore, the lateral area of the cylinder is found by multiplying its height by the circumference of its base. Written as a formula, this is

$$A = Ch$$

EXAMPLE: Find the lateral area of a right circular cylinder whose base has a radius of 4 inches and whose height is 6 inches.
SOLUTION: The circumference of the base is

$$C = \pi d$$
$$C = 3.14 \text{ x } 8 \text{ in.}$$
$$= 25.12 \text{ in.}$$

Therefore,

$$A = 25.12 \text{ in. x } 6 \text{ in.}$$
$$= 151 \text{ sq in. (approximately)}$$

The formula for the volume of a cylinder is obtained by the same reasoning process that was used for prisms. The cylinder is considered to be composed of many circular wafers, or disks, each one unit thick. The area of each disk, multiplied by the number of disks, is the volume of the cylinder. With V representing volume, A representing the area of each disk, and n representing the number of disks, the formula is as follows:

$$V = An$$

Since the number of disks is the same as the height of the cylinder, the formula for the volume of a cylinder is normally written

$$V = Bh$$

In this formula, B is the area of the base and h is the height of the cylinder.

EXAMPLE: Determine the volume of a circular cylinder with a base of radius 5 inches and a height of 14 inches.

SOLUTION:

$$
\begin{aligned}
V &= Bh \\
&= (\pi \times 5^2) \times 14 \\
&= \frac{22}{7} \times 25 \times 14 \\
&= 22 \times 25 \times 2 \\
&= 22 \times 50 \\
&= 1,100 \text{ cu in.}
\end{aligned}
$$

Practice problems:

1. Determine the lateral area of a right circular cylinder with a base of diameter 7 inches and a height of 4 inches.

2. Determine the volume of the cylinder in problem 1.

Answers:

1. 88 sq in.   2. 154 cu in.

REGULAR PYRAMIDS AND RIGHT CIRCULAR CONES

A PYRAMID is a solid figure, the lateral faces of which are triangles. (See fig. 18-16.) A REGULAR PYRAMID has all of its lateral faces equal.

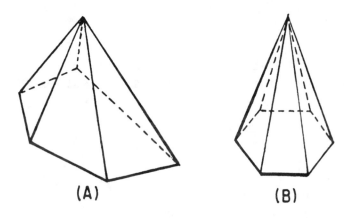

(A)   (B)

Figure 18-16.—(A) Irregular pyramid; (B) regular pyramid.

A regular pyramid with a very large number of lateral faces would have a base polygon with many sides. If the number of sides is sufficiently large, the base polygon is indistinguishable from a circle and the surface formed by the many lateral faces becomes a smoothly curved surface. The solid figure thus formed is a RIGHT CIRCULAR CONE. (See fig. 18-17.)

530

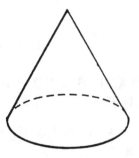

Figure 18-17.—Right circular cone.

Slant Height

The slant height of a regular pyramid is the perpendicular distance from the vertex to the center of any side of the base. For example,

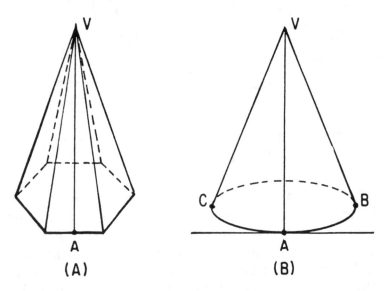

Figure 18-18.—(A) Slant height of a regular pyramid; (B) slant height of a right circular cone.

the length of line AV in figure 18-18 (A) is the slant height. The slant height of a right circular cone is the length of any straight line joining the vertex to the circumference line of the base. Such a line is perpendicular to a line tangent to the base at the point where the slant height intersects the base. (See fig. 18-18 (B).) Lines AV, BV, and CV in figure 18-18 (B) are all slant heights.

Lateral Area

The lateral area of a pyramid is the sum of the areas of its lateral faces. If the pyramid is regular, its lateral faces have equal bases; furthermore, the slant height is the altitude of each face. Therefore, the area of each lateral face is one-half the slant height multiplied by the length of one side of the base polygon. Since the sum of these sides is the perimeter of the base, the total lateral area of the pyramid is the product of one-half its slant height multiplied by the perimeter of its base. Using s to represent slant height and P to represent the perimeter of the base, the formula is as follows:

$$\text{Lateral Area} = \frac{1}{2} sP$$

A right circular cone can be considered as a regular pyramid with an infinite number of faces. Therefore, using C to represent the circumference of the base, the formula for the lateral area of a right circular cone is

$$\text{Lateral Area} = \frac{1}{2} sC$$

Volume

The volume of a pyramid is determined by its base and its altitude, as is the case with other solid figures. Experiments show that the volume of any pyramid is one-third of the product of its base and its altitude. This may be stated as a formula with V representing volume, B representing the area of the base, and h representing height (altitude), as follows:

$$V = \frac{1}{3} Bh$$

The formula for the volume of a pyramid does not depend in any way upon the number of faces. Therefore, we use the same formula for the volume of a right circular cone. Since the base is a circle, we replace B with πr^2 (where r is the radius of the base). The formula for the volume of a right circular cone is then

$$V = \frac{1}{3} Bh$$

$$= \frac{1}{3} \pi r^2 h$$

Practice problems:

1. Find the lateral area of a regular pyramid with a 5-sided base measuring 3 inches on each side, if the slant height is 12 inches.

2. Find the lateral area of a right circular cone whose base has a diameter of 6 cm and whose slant height is 14 cm.

3. Find the volume of a regular pyramid with a square base measuring 4 cm on each side, if the vertex is 9 cm above the base.

4. Find the volume of a right circular cone whose base has a diameter of 14 inches, if the altitude is 21 inches.

Answers:

1. 90 sq in.   3. 48 cu cm

2. 132 sq cm   4. 1078 cu in.

SPHERES

A SPHERE is a solid figure with all points on its surface equally distant from its center.

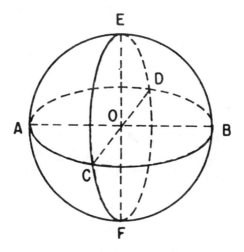

Figure 18-19.—Parts of a sphere.

In figure 18-19, the center of the sphere is point O.

A RADIUS of a sphere is a straight line segment joining the center of the sphere to a point on the surface. Lines OA, OB, OC, OD, OE, and OF in figure 18-19 are radii. A DIAMETER of a sphere is a straight line segment joining two points on the surface and passing through the center of the sphere. Lines AB, CD, and EF in figure 18-19 are diameters. A HEMISPHERE is half of a sphere.

Circles of various sizes may be drawn on the surface of a sphere. The largest circle that may be so drawn is one with a radius equal to the radius of the sphere. Such a circle is a GREAT CIRCLE. In figure 18-19, circles AEBF, ACBD, and CEDF are great circles.

On the surface of a sphere, the shortest distance between two points is an arc of a great circle drawn so that it passes through the two points. This explains the importance of great circles in the science of navigation, since the earth is approximately a sphere.

Surface Area

The surface area of a sphere may be calculated by multiplying 4 times π times the square of the radius. Written as a formula, this is

$$A = 4\pi r^2$$

The formula for the surface area of a sphere may be rewritten as follows:

$$A = (2\pi r)(2r)$$

When the formula is factored in this way, it is easy to see that the surface area of a sphere is simply its circumference times its diameter.

Volume

The volume of a sphere whose radius is r is given by the formula

$$V = \frac{4}{3} \pi r^3$$

EXAMPLE: Find the volume of a sphere whose diameter is 42 inches.

SOLUTION:

$$V = \frac{4}{3} \pi r^3$$

$$= \frac{4}{3} \times \frac{22}{7} \times (21 \text{ in.})^3$$

$$= \frac{4}{3} \times \frac{22}{7} \times 21 \times 21 \times 21 \text{ cu in.}$$

$$= 88 \times 21 \times 21 \text{ cu in.}$$

$$= 38,808 \text{ cu in.}$$

Practice problems. Calculate the surface area and the volume of the sphere in each of the following problems:

1. Radius = 7 inches   2. Radius = 14 cm

Answers:

1. Area = 616 sq in.
   Volume = 1,437 cu in. (approx.)

2. Area = 2,464 sq cm
   Volume = 11,499 cu cm (approx.)

CHAPTER 19

NUMERICAL TRIGONOMETRY

The word "trigonometry" means "measurement by triangles." As it is presented in many textbooks, trigonometry includes topics other than triangles and measurement. However, this chapter is intended only as an introduction to the numerical aspects of trigonometry as they relate to measurement of lengths and angles.

SPECIAL PROPERTIES OF RIGHT TRIANGLES

A RIGHT TRIANGLE has been defined as any triangle containing a right angle. The side opposite the right angle in a right triangle is a HYPOTENUSE. (See fig. 19-1.) In figure 19-1, side AC is the hypotenuse.

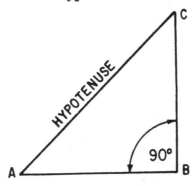

Figure 19-1.—A right triangle.

An important property of all right triangles, which relates the lengths of the three sides, was discovered by the Greek philosopher Pythagoras.

PYTHAGOREAN THEOREM

The rule of Pythagoras, or PYTHAGOREAN THEOREM, states that the square of the length of the hypotenuse (in any right triangle) is equal to the sum of the squares of the lengths of the other two sides. For example, if the sides are

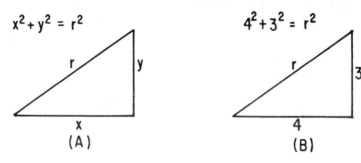

Figure 19-2.—The Pythagorean Theorem. (A) General triangle; (B) triangle with sides of specific lengths.

labeled as in figure 19-2 (A), the Pythagorean Theorem is stated in symbols as follows:

$$x^2 + y^2 = r^2$$

An example of the use of the Pythagorean Theorem in a problem follows:

EXAMPLE: Find the length of the hypotenuse in the triangle shown in figure 19-2 (B).

538

SOLUTION: $r^2 = 3^2 + 4^2$

$$r = \sqrt{9 + 16}$$

$$= \sqrt{25} = 5$$

EXAMPLE: An observer on a ship at point A, figure 19-3, knows that his distance from point C is 1,200 yards and that the length of BC is 1,300 yards. He measures angle A and finds that it is 90°. Calculate the distance from A to B.

SOLUTION: By the rule of Pythagoras,

$$(BC)^2 = (AB)^2 + (AC)^2$$
$$(1,300)^2 = (AB)^2 + (1,200)^2$$
$$(1,300)^2 - (1,200)^2 = (AB)^2$$

$$(13 \times 10^2)^2 - (12 \times 10^2)^2 = (AB)^2$$
$$(169 \times 10^4) - (144 \times 10^4) = (AB)^2$$

$$25 \times 10^4 = (AB)^2$$

$$500 \text{ yd} = AB$$

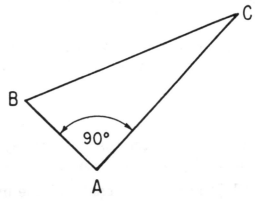

Figure 19-3.—Using the Pythagorean Theorem.

SIMILAR RIGHT TRIANGLES

Two right triangles are SIMILAR if one of the acute angles of the first is equal to one of the acute angles of the second. This conclusion is supported by the following reasons:

1. The right angle in the first triangle is equal to the right angle in the second, since all right angles are equal.

2. The sum of the angles of any triangle is 180°. Therefore, the sum of the two acute angles in a right triangle is 90°.

3. Let the equal acute angles in the two triangles be represented by A and A' respectively. (See fig. 19-4.) Then the other acute angles, B and B', are as follows:

$$B = 90° - A$$
$$B' = 90° - A'$$

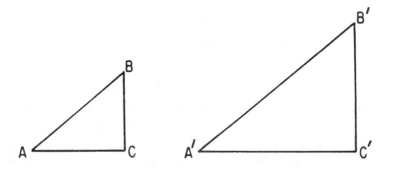

Figure 19-4.—Similar right triangles.

4. Since angles A and A' are equal, angles B and B' are also equal.

5. We conclude that two right triangles with one acute angle of the first equal to one acute

angle of the second have all of their corresponding angles equal. Thus the two triangles are similar.

Practical situations frequently occur in which similar right triangles are used to solve problems. For example, the height of a tree can be determined by comparing the length of its shadow with that of a nearby flagpole, as shown in figure 19-5.

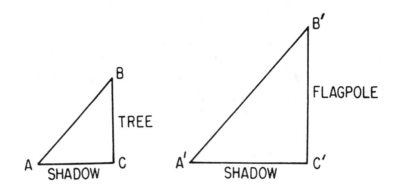

Figure 19-5.—Calculation of height by comparison of shadows.

Assume that the rays of the sun are parallel and that the tree and flagpole both form 90° angles with the ground. Then triangles ABC and A'B'C' are right triangles and angle B is equal to angle B'. Therefore, the triangles are similar and their corresponding sides are proportional, with the following result:

$$\frac{BC}{AC} = \frac{B'C'}{A'C'}$$

$$BC = \frac{(AC) \times (B'C')}{A'C'}$$

541

Suppose that the flagpole is known to be 30 feet high, the shadow of the tree is 12 feet long, and the shadow of the flagpole is 24 feet long. Then

$$BC = \frac{12 \times 30}{24} = 15 \text{ feet}$$

Practice problems.

1. A mast at the top of a building casts a shadow whose tip is 48 feet from the base of the building. If the building is 12 feet high and its shadow is 32 feet long, what is the length of the mast? (NOTE: If the length of the mast is x, then the height of the mast above the ground is x + 12.)

2. Figure 19-6 represents an L-shaped building with dimensions as shown. On the line of sight from A to D, a stake is driven at C, a point 8 feet from the building and 10 feet from A. If ABC is a right angle, find the length of AB and the length of AD. Notice that AE is 18 feet and ED is 24 feet.

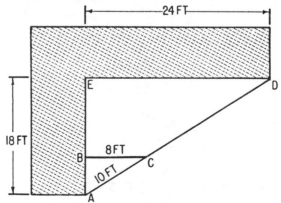

Figure 19-6.—Using similar triangles.

542

Answers:

1. 6 feet

2. AB = 6 feet

   AD = 30 feet

TRIGONOMETRIC RATIOS

The relationships between the angles and the sides of a right triangle are expressed in terms of TRIGONOMETRIC RATIOS. For example, in figure 19-7, the sides of the triangle are named in accordance with their relationship to angle θ. In trigonometry, angles are usually named by means of Greek letters. The Greek name of the symbol θ is theta.

The six trigonometric ratios for the angle θ are listed in table 19-1.

The ratios are defined as follows:

1. $\sin \theta = \dfrac{\text{side opposite } \theta}{\text{hypotenuse}} = \dfrac{y}{r}$

2. $\cos \theta = \dfrac{\text{side adjacent to } \theta}{\text{hypotenuse}} = \dfrac{x}{r}$

3. $\tan \theta = \dfrac{\text{side opposite } \theta}{\text{side adjacent to } \theta} = \dfrac{y}{x}$

4. $\cot \theta = \dfrac{\text{side adjacent to } \theta}{\text{side opposite } \theta} = \dfrac{x}{y}$

5. $\sec \theta = \dfrac{\text{hypotenuse}}{\text{side adjacent to } \theta} = \dfrac{r}{x}$

6. $\csc \theta = \dfrac{\text{hypotenuse}}{\text{side opposite to } \theta} = \dfrac{r}{y}$

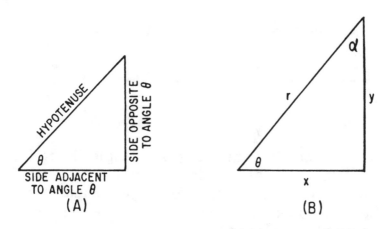

Figure 19-7.—Relationship of sides and angles
in a right triangle. (A) Names of the sides;
(B) symbols used to designate the sides.

Table 19-1.—Trigonometric ratios.

Name of ratio	Abbreviation
sine of θ	sin θ
cosine of θ	cos θ
tangent of θ	tan θ
cotangent of θ	cot θ
secant of θ	sec θ
cosecant of θ	csc θ

The other acute angle in figure 19-7 (B) is
labeled α (Greek alpha). The side opposite α
is x and the side adjacent to α is y. Therefore
the six ratios for α are as follows:

544

1. $\sin \alpha = \dfrac{x}{r}$

4. $\cot \alpha = \dfrac{y}{x}$

2. $\cos \alpha = \dfrac{y}{r}$

5. $\sec \alpha = \dfrac{r}{y}$

3. $\tan \alpha = \dfrac{x}{y}$

6. $\csc \alpha = \dfrac{r}{x}$

Suppose that the sides of triangle (B) in figure 19-7 are as follows: x = 3, y = 4, r = 5. Then each of the ratios for angles θ and α may be expressed as a common fraction or as a decimal. For example,

$$\sin \theta = \frac{4}{5} = 0.800$$

$$\sin \alpha = \frac{3}{5} = 0.600$$

Decimal values have been computed for ratios of angles between 0° and 90°, and values for angles above 90° can be expressed in terms of these same values by means of conversion formulas. Appendix II of this training course gives the sine, cosine, and tangent of angles from 0° to 90°. The secant, cosecant, and cotangent are calculated, when needed, by using their relationships to the three principal ratios. These relationships are as follows:

$$\text{secant } \theta = \frac{1}{\text{cosine } \theta}$$

$$\text{cosecant } \theta = \frac{1}{\text{sine } \theta}$$

$$\text{cotangent } \theta = \frac{1}{\text{tangent } \theta}$$

TABLES

Tables of decimal values for the trigonometric ratios may be constructed in a variety
of ways. Some give the angles in degrees, minutes, and seconds; others in degrees and tenths
of a degree. The latter method is more com-
pact and is the method used for appendix II.
The "headings" at the bottom of each page in
appendix II provide a convenient reference
showing the minute equivelents for the decimal
fractions of a degree. For example, 12' (12
minutes) is the equivalent of 0.2°.

Finding the Function Value

The trigonometric ratios are sometimes
called FUNCTIONS, because the value of the
ratio depends upon (is a function of) the angle
size. Finding the function value in appendix II
is easily accomplished. For example, the sine
35° is found by looking in the "sin" row opposite the large number 35, which is located in
the extreme left-hand column.

Since our angle in this example is exactly
35°, we look for the decimal value of the sine
in the column with the 0.0° heading. This column contains decimal values for functions of
the angle plus 0.0°; in our example, 35° plus
0.0°, or simply 34.0°. Thus we find that the
sine of 35.0° is 0.5736. By the same reasoning,
the sine of 42.7° is 0.6782, and the tangent of
32.3° is 0.6322.

A typical problem in trigonometry is to find
the value of an unknown side in a right triangle
when only one side and one acute angle are

known. EXAMPLE: In triangle ABC (fig. 19-8), find the length of AC if AB is 13 units long and angle CAB is 34.7°.

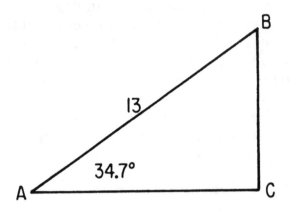

Figure 19-8.—Using the trigonometric ratios to evaluate the sides.

SOLUTION:

$$\frac{AC}{13} = \cos 34.7°$$

$$AC = 13 \cos 34.7°$$

$$= 13 \times 0.8221$$

$$= 10.69 \text{ (approx.)}$$

The angles of a triangle are frequently stated in degrees and minutes, rather than degrees and tenths. For example, in the foregoing problem, the angle might have been stated as 34°42'. When the stated number of minutes is an exact multiple of 6 minutes, the minute entries at the bottom of each page in appendix II may be used.

Finding the Angle

Problems are frequently encountered in which two sides are known, in a right triangle, but neither of the acute angles is known. For example, by applying the Pythagorean Theorem we can verify that the triangle in figure 19-9 is

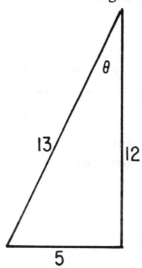

Figure 19-9.—Using trigonometric ratios to evaluate angles.

a right triangle. The only information given, concerning angle θ, is the ratio of sides in the triangle. The size of θ is calculated as follows:

$$\tan \theta = \frac{5}{12} = 0.4167$$

θ = the angle whose tangent is 0.4167

Assuming that the sides and angles in figure 19-9 are in approximately the correct proportions, we estimate that angle θ is about $20°$. The table entries for the tangent in the vicinity of $20°$ are slightly too small, since we need a number near 0.4167. However, the tangent of $22°36'$ is 0.4163 and the tangent of $22°42'$ is 0.4183. Therefore, θ is between $22°36'$ and $22°42'$.

Interpolation

It is frequently necessary to estimate the value of an angle to a closer approximation than is available in the table. This is equivalent to estimating between table entries, and the process is called INTERPOLATION. For example, in the foregoing problem it was determined that the angle value was between $22°36'$ and $22°42'$. The following paragraphs describe the procedure for interpolating to find a closer approximation to the value of the angle.

The following arrangement of numbers is recommended for interpolation:

ANGLE	TANGENT

$$6' \left\{ \begin{array}{ll} 22°36' & 0.4163 \\ \theta & 0.4167 \\ 22°42' & 0.4183 \end{array} \right. \quad \left. \begin{array}{c} \\ \end{array} \right\} .0004 \quad \left. \begin{array}{c} \\ \\ \end{array} \right\} .0020$$

The spread between $22°36'$ and $22°42'$ is $6'$, and we use the comparison of the tangent values to determine how much of this $6'$ spread is included in θ, the angle whose value is sought. Notice that the tangent of θ is different from

549

tan 22°36' by only 0.0004, and the total spread in the tangent values is 0.0020. Therefore, the tangent of θ is $\dfrac{0.0004}{0.0020}$ of the way between the tangents of the two angles given in the table. This is 1/5 of the total spread, since

$$\frac{0.0004}{0.0020} = \frac{4}{20} = \frac{1}{5}$$

Another way of arriving at this result is to observe that the total spread is 20 ten-thousandths, and that the partial spread corresponding to angle θ is 4 ten-thousandths. Since 4 out of 20 is the same as 1 out of 5, angle θ is 1/5 of the way between 22°36' and 22°42'.

Taking 1/5 of the 6' spread between the angles, we have the following calculation:

$$\frac{1}{5} \times 6' = \frac{1}{5} \times 5'60''$$

$$= 1'12'' \text{ (1 minute and 12 seconds)}$$

The 12" obtained in this calculation causes our answer to appear to have greater accuracy than the tables from which it is derived. This apparent increase in accuracy is a normal result of interpolation. Final answers based on interpolated data should be rounded off to the same degree of accuracy as that of the original data.

The value of 1 minute and 12 seconds found in the foregoing problem is added to 22° 36', as follows:

$$\theta = 22°36' + 1'12'' = 22°37'12''$$

Therefore θ is 22°37', approximately.

The foregoing problem could have been solved in terms of tenths and hundredths of a degree, rather than minutes, as follows:

ANGLE   TANGENT

$$0.1° \begin{cases} 22.60° & 0.4163 \\ \theta & 0.4167 \\ 22.70° & 0.4183 \end{cases}$$

with brackets showing 0.0004, 0.0020

In this example, we are concerned with an angular spread of 0.10° and θ is located 1/5 of the way through this spread. Thus we have

$$\theta = 22.60° + \left(\frac{1}{5} \times 0.10°\right)$$

$$\theta = 22.60° + 0.02°$$

$$\theta = 22.62°$$

Interpolation must be approached with commonsense, in order to avoid applying corrections in the wrong direction. For example, the cosine of an angle decreases in value as the angle increases from 0° to 90°. If we need the value of the cosine of an angle such as 22°39', the calculation is as follows:

ANGLE   COSINE

$$6' \begin{cases} 22°36' & 0.9232 \\ 22°39' & \\ 22°42' & 0.9225 \end{cases}$$

with brackets showing 3', 0.0007

In this example, it is easy to see that 22°39' is halfway between 22°36' and 22°42'. Therefore the cosine of 22°39' is halfway between the cosine of 22°36' and that of 22°42'. Taking one-half of the spread between these cosines, we then SUBTRACT from 0.9232 to find the cosine of 22°39', as follows:

$$\cos 22°39' = 0.9232 - \left(\frac{1}{2} \times 0.0007\right)$$

$$= 0.9232 - 0.00035$$

$$= 0.92285$$

$$= 0.9229 \text{ (approximately)}$$

Practice problems:

1. Use the table in appendix II to find the decimal value of each of the following ratios:

a. tan 45°

b. sin 60°

c. cos 42°6'

d. sin 37°14'

e. cos 51.5°

f. tan 13.75°

2. Find the angle which corresponds to each of the following decimal values in appendix II:

a. sin θ = 0.2790

b. cos θ = 0.9018

c. tan θ = 0.7604

d. sin θ = 0.8142

Answers:

1. a. 1

b. 0.8660

c. 0.7420

d. 0.6051

e. 0.6225

f. 0.2447

2. a. $\theta = 16.2°$ c. $\theta = 37°15'$

   b. $\theta = 25°36'$ d. $\theta = 54°30'$

RIGHT TRIANGLES WITH SPECIAL ANGLES AND SIDE RATIOS

Three types of right triangles are expecially significant because of their frequent occurrence. These are the 30°-60°-90° triangle, the 45°-90° triangle, and the 3-4-5 triangle.

THE 30° - 60° - 90° TRIANGLE

The 30°-60°-90° triangle is so named because these are the sizes of its three angles. The sides of this triangle are in the ratio of 1 to √3 to 2, as shown in figure 19-10.

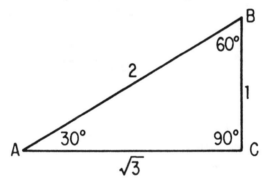

Figure 19-10.—30°-60°-90° triangle.

The sine ratio for the 30° angle in figure 19-10 establishes the proportionate values of the sides. For example, we know that the sine of 30° is 1/2; therefore side AB must be twice as long as BC. If side BC is 1 unit long, then side AB is 2 units long and, by the rule of Pythagoras, AC is found as follows:

553

$$AC = \sqrt{(AB)^2 - (BC)^2}$$
$$= \sqrt{4 - 1} = \sqrt{3}$$

Regardless of the size of the unit, a 30°-60°-90° triangle has a hypotenuse which is 2 times as long as the shortest side. The shortest side is opposite the 30° angle. The side opposite the 60° angle is $\sqrt{3}$ times as long as the shortest side. For example, suppose that the hypotenuse of a 30°-60°-90° triangle is 30 units long; then the shortest side is 15 units long, and the length of the side opposite the 60° angle is 15 $\sqrt{3}$ units.

Practice problems. Without reference to tables or to the rule of Pythagoras, find the following lengths and angles in figure 19-11:

1. Length of AC.
2. Size of angle A.
3. Size of angle B.
4. Length of RT.
5. Length of RS.
6. Size of angle T.

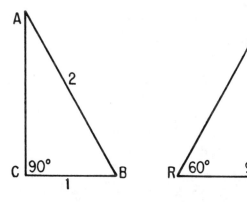

Figure 19-11.—Finding parts of 30°-60°-90° triangles.

Answers:

1. $\sqrt{3}$ 4. 4

2. 30°   5. 2

3. 60°   6. 30°

THE 45° - 90° TRIANGLE

Figure 19-12 illustrates a triangle in which two angles measure 45° and the third angle

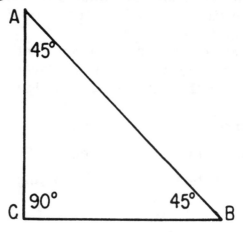

Figure 19-12.—A 45°-90° triangle.

measures 90°. Since angles A and B are equal, the sides opposite them are also equal. Therefore, AC equals CB. Suppose that CB is 1 unit long; then AC is also 1 unit long, and the length of AB is calculated as follows:

$$(AB)^2 = 1^2 + 1^2 = 2$$
$$AB = \sqrt{2}$$

Regardless of the size of the triangle, if it has two 45° angles and one 90° angle, its sides are in the ratio 1 to 1 to $\sqrt{2}$. For example, if sides AC and CB are 3 units long, AB is 3 $\sqrt{2}$ units long.

Practice problems. Without reference to tables or to the rule of Pythagoras, find the following lengths and angles in figure 19-13:

1. AB   2. BC   3. Angle B

Answers:

1. 2 $\sqrt{2}$ 2. 2   3. 45°

THE 3-4-5 TRIANGLE

The triangle shown in figure 19-14 has its sides in the ratio 3 to 4 to 5. Any triangle with its sides in this ratio is a right triangle.

It is a common error to assume that a triangle is a 3-4-5 type because two sides are known to be in the ratio 3 to 4, or perhaps 4 to 5. Figure 19-15 shows two examples of triangles which happen to have two of their sides in the stated ratio, but not the third side. This can be because the triangle is not a right triangle, as in figure 19-15 (A). On the other hand, even though the triangle is a right triangle its longest side may be the 4-unit side, in which case the third side cannot be 5 units long. (See fig. 19-15 (B).)

It is interesting to note that the third side in figure 19-15 (B) is $\sqrt{7}$. This is a very unusual coincidence, in which one side of a right tri-

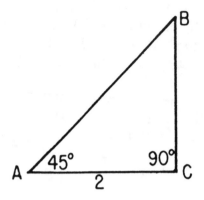

Figure 19-13.—Finding unknown parts
in a 45°-90° triangle.

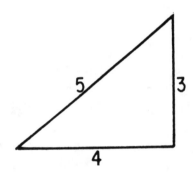

Figure 19-14.—A 3-4-5 triangle.

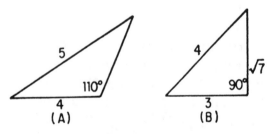

Figure 19-15.—Triangles which may be
mistaken for 3-4-5 triangles.

angle is the square root of the sum of the other two sides.

Related to the basic 3-4-5 triangle are all triangles whose sides are in the ratio 3 to 4 to 5 but are longer (proportionately) than these basic lengths. For example, the triangle pictured in figure 19-6 is a 3-4-5 triangle.

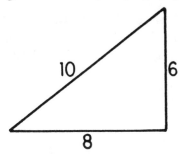

Figure 19-16.—Triangle with sides which are multiples of 3, 4, and 5.

The 3-4-5 triangle is very useful in calculations of distance. If the data can be adapted to fit a 3-4-5 configuration, no tables or calculation of square root (Pythagorean Theorem) are needed.

EXAMPLE: An observer at the top of a 40-foot vertical tower knows that the base of the tower is 30 feet from a target on the ground. How does he calculate his slant range (direct line of sight) from the target?

SOLUTION: Figure 19-17 shows that the desired length, AB, is the hypotenuse of a right triangle whose shorter sides are 30 feet and 40 feet long. Since these sides are in the ratio 3 to 4 and angle C is 90°, the triangle is a 3-4-5

558

triangle. Therefore, side AB represents the 5-unit side of the triangle. The ratio 30 to 40 to 50 is equivalent to 3-4-5, and thus side AB is 50 units long.

Practice problems. Without reference to tables or to the rule of Pythagoras, solve the following problems:

1. An observer is at the top of a 30-foot vertical tower. Calculate his slant range from a target on the ground which is 40 feet from the base of the tower.

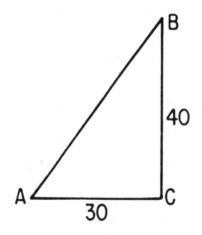

Figure 19-17.—Solving problems with a 3-4-5 triangle.

2. A guy wire 15 feet long is stretched from the top of a pole to a point on the ground 9 feet from the base of the pole. Calculate the height of the pole.

Answers:

1. 50 feet   2. 12 feet

OBLIQUE TRIANGLES

Oblique triangles were defined in chapter 17 of this training course as triangles which contain no right angles. A natural approach to the solution of problems involving oblique triangles is to construct perpendicular lines and form right triangles which subdivide the original triangle. Then the problem is solved by the usual methods for right triangles.

DIVISION INTO RIGHT TRIANGLES

The oblique triangle ABC in figure 19-18 has been divided into two right triangles by drawing line BD perpendicular to AC. The length of AC is found as follows:

1. Find the length of AD.

$$\frac{AD}{35} = \cos 40°$$

$$AD = 35 \cos 40°$$

$$= 35 \ (0.7660)$$

$$= 26.8 \ (\text{approximately})$$

CAUTION: A careless appraisal of this problem may lead the unwary trainee to represent the ratio AC/AB as the cosine of 40°. This error is avoided only by the realization that the trigonometric ratios are based on RIGHT triangles.

2. In order to find the length of DC, first calculate BD.

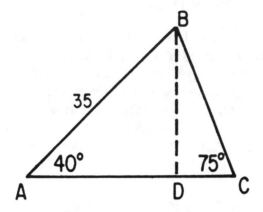

Figure 19-18.—Finding the unknown parts
of an oblique triangle.

$$\frac{BD}{35} = \sin 40°$$

$$BD = 35 \sin 40°$$

$$= 35 (0.6428)$$

$$= 22.4 \text{ (approximately)}$$

3. Find the length of DC.

$$\frac{22.4}{DC} = \tan 75°$$

$$DC = \frac{22.4}{\tan 75°} = \frac{22.4}{3.732}$$

$$DC = 6.01 \text{ (approximately)}$$

4. Add AD and DC to find AC.

$$26.8 + 6.01 = 32.81$$

$$AC = 32.8 \text{ (approximately)}$$

SOLUTION BY SIMULTANEOUS EQUATIONS

A typical problem in trigonometry is the determination of the height of a point such as B in figure 19-19.

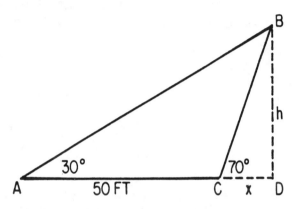

Figure 19-19.—Calculation of unknown quantities by means of oblique triangles.

Suppose that point B is the top of a hill, and point D is inaccessible. Then the only measurements possible on the ground are those shown in figure 19-19. If we let h represent BD and x represent CD, we can set up the following system of simultaneous equations:

$$\frac{h}{x} = \tan 70°$$

$$\frac{h}{50 + x} = \tan 30°$$

Solving these two equations for h in terms of x, we have

562

$$h = x \tan 70°$$

and

$$h = (50 + x) \tan 30°$$

Since the two quantities which are both equal to h must be equal to each other, we have

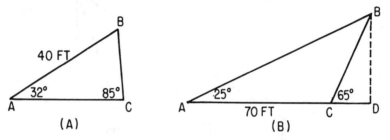

Figure 19-20.—(A) Oblique triangle with all angles acute; (B) obtuse triangle.

$$x \tan 70° = (50 + x) \tan 30°$$

$$x (2.748) = 50 (0.5774) + x(0.5774)$$

$$x (2.748) - x (0.5774) = 28.8$$

$$x (2.171) = 28.8$$

$$x = \frac{28.8}{2.171} = 13.3 \text{ feet}$$

Knowing the value of x, it is now possible to compute h as follows:

$$h = x \tan 70°$$

$$= 13.3 (2.748)$$

$$= 36.5 \text{ feet (approximately)}$$

Practice problems:

1. Find the length of side BC in figure 19-20 (A).

2. Find the height of point B above line AD in figure 19-20 (B).

Answers:

1. 21.3 feet   2. 41.7 feet

LAW OF SINES

The law of sines provides a direct approach to the solution of oblique triangles, avoiding the necessity of subdividing into right triangles. Let the triangle in figure 19-21 (A) represent any oblique triangle with all of its angles acute.

The labels used in figure 19-21 are standardized. The small letter a is used for the side opposite angle A; small b is opposite angle B; small c is opposite angle C.

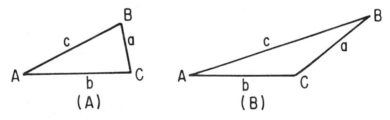

Figure 19-21.—(A) Acute oblique triangle with standard labels; (B) obtuse triangle with standard labels.

The law of sines states that in any triangle, whether it is acute as in figure 19-21 (A) or obtuse as in figure 19-21 (B), the following is true:

$$\frac{a}{\sin\ A} = \frac{b}{\sin\ B} = \frac{c}{\sin\ C}$$

EXAMPLE: In figure 19-21 (A), let angle A be 15° and let angle C be 85°. If BC is 20 units, find the length of AB.

SOLUTION: By the law of sines,

$$\frac{20}{\sin\ 15°} = \frac{c}{\sin\ 85°}$$

$$c = \frac{20\ \sin\ 85°}{\sin\ 15°}$$

$$c = \frac{20\ (0.9962)}{0.2588} = 77.0$$

CHAPTER 20

A REVIEW ON SOLVING TRIANGLES

LAWS AND FORMULAS

The following results hold for any oblique triangle ABC with sides of length a,b,c opposite to vertices A,B,C respectively.

Law of Sines

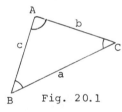

Fig. 20.1

$$\frac{a}{\sin A} = \frac{b}{\sin B} = \frac{c}{\sin C}$$

Law of Cosines

$$a^2 = b^2 + c^2 - 2bc \cos A.$$

$$b^2 = a^2 + c^2 - 2ac \cos B.$$

$$c^2 = a^2 + b^2 - 2ab \cos C.$$

Law of Tangents

$$\frac{a - b}{a + b} = \frac{\tan\left(\dfrac{A - B}{2}\right)}{\tan\left(\dfrac{A + B}{2}\right)}$$

$$\frac{b - c}{b + c} = \frac{\tan\left(\dfrac{B - C}{2}\right)}{\tan\left(\dfrac{B + C}{2}\right)}$$

$$\frac{a - c}{a + c} = \frac{\tan\left(\dfrac{A - C}{2}\right)}{\tan\left(\dfrac{A + C}{2}\right)}$$

Mollweide's Formulas

$$\frac{a + b}{c} = \frac{\cos \frac{1}{2} (A - B)}{\sin \frac{C}{2}}, \qquad \frac{a - b}{c} = \frac{\sin \frac{1}{2} (A - B)}{\cos \frac{C}{2}}$$

$$\frac{b + c}{a} = \frac{\cos \frac{1}{2} (B - C)}{\sin \frac{A}{2}}, \qquad \frac{b - c}{a} = \frac{\sin \frac{1}{2} (B - C)}{\cos \frac{A}{2}}$$

$$\frac{c + a}{b} = \frac{\cos \frac{1}{2} (C - A)}{\sin \frac{A}{2}}, \qquad \frac{c - a}{b} = \frac{\sin \frac{1}{2} (C - A)}{\cos \frac{B}{2}}$$

Projection Formulas

$$BC = b \cos C + c \cos B$$

$$AC = a \cos C + c \cos A$$

$$AB = a \cos B + b \cos A$$

In a 30°-60° right triangle, (Fig. 20.2), the hypotenuse is twice the length of the side opposite the 30° angle. The side opposite the 60° angle is equal to the length of the side opposite the 30° angle multiplied by $\sqrt{3}$.

In an isosceles 45° right triangle, (Fig. 20.3), the hypotenuse is equal to the length of one of its arms multiplied by $\sqrt{2}$.

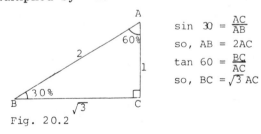

$\sin 30 = \frac{AC}{AB}$

so, $AB = 2AC$

$\tan 60 = \frac{BC}{AC}$

so, $BC = \sqrt{3}\ AC$

$$\sin 30° = \frac{AC}{AB}$$

$$\text{so, } AB = 2AC$$

$$\tan 60° = \frac{BC}{AC}$$

Fig. 20.2

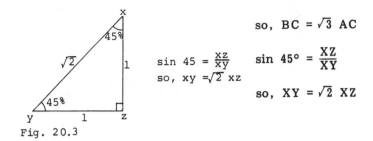

so, BC = $\sqrt{3}$ AC

$\sin 45 = \frac{XZ}{XY}$ $\sin 45° = \frac{XZ}{XY}$

so, xy = $\sqrt{2}$ xz

so, XY = $\sqrt{2}$ XZ

Fig. 20.3

IMPORTANT CONCEPTS AND THEOREMS

The altitude h on the hypotenuse of a right triangle is the mean proportional between the segments of the hypotenuse, also called the geometric mean.

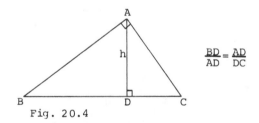

$$\frac{BD}{AD} = \frac{AD}{DC}$$

Fig. 20.4

In a right triangle △ ABC, the altitude to the hypotenuse, $\overline{AD}$, separates the triangle into two triangles that are similar to each other and to the original triangle.

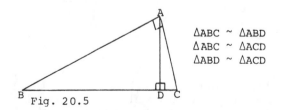

△ABC ~ △ABD
△ABC ~ △ACD
△ABD ~ △ACD

Fig. 20.5

The length of the median to the hypotenuse of a right

triangle is equal to one-half the length of the hypotenuse.

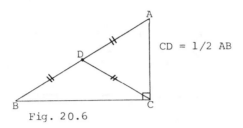

CD = 1/2 AB

Fig. 20.6

In solving triangles, terms such as line of sight, angle of elevation and angle of depression are often used. These terms are illustrated below.

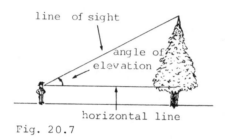

line of sight

angle of elevation

horizontal line

Fig. 20.7

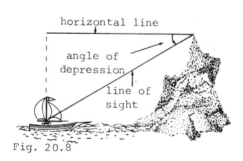

horizontal line

angle of depression

line of sight

Fig. 20.8

Ex. At a point on the ground 40 feet from the foot of a tree, the angle of elevation to the top of the tree is 42°. Find the height of the tree to the nearest foot.

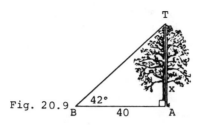

Fig. 20.9

Solution: The geometric figure formed is a right triangle (see figure). Since the unknown height of the tree is opposite the given angle of elevation, and we are given the side adjacent to this angle, we can solve the problem using the tangent ratio. The tangent is the ratio of the length of the leg opposite the acute angle to the length of the leg adjacent to the acute angle in any right triangle. In this example,

$$\tan B = \frac{\text{length of leg opposite} \quad \angle B}{\text{lenght of leg adjacent} \quad \angle B},$$

$$\tan B = \frac{AT}{BA}.$$

Let $x = AT$, and consult a standard table of tangents to find that $\tan 42° = 0.9004$. Since $BA = 40$, we obtain

$$0.9004 = \frac{x}{40}.$$

Therefore, $x = 40(0.9004) = 36.016$.

Therefore, the height of the tree, to the nearest foot, is 36 feet.

The following theorems are often used for finding the area of a triangle.

Theorem: The area of a triangle is given by $A = \frac{1}{2}bh$, where b is the length of the base and h is the perpendicular height of the triangle.

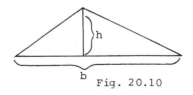

Fig. 20.10

570

Theorem: The area of a triangle equals one-half the product of any two adjacent sides and the sine of the included angle.

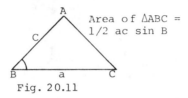

Area of $\triangle ABC =$
$1/2$ ac sin B

Fig. 20.11

Theorem: Triangles that share the same base and have their third vertex on a line parallel to the base, have equal areas.

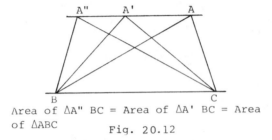

Area of $\triangle A''$ BC = Area of $\triangle A'$ BC = Area of $\triangle ABC$

Fig. 20.12

The areas of two triangles having equal bases, have the same ratio as that of their altitudes and vice versa.

The area of a triangle, the length of whose three sides are a, b, and c, is given by the formula

$$A = \sqrt{s(s - a)(s - b)(s - c)}$$

where $s = \frac{1}{2}(a + b + c)$; the semiperimeter of the triangle. The above formula is commonly referred to as Heron's formula.

The area of an equilateral triangle is given by the formula,

$$A = \frac{x^2 \sqrt{3}}{4},$$

where x is the length of a side of the triangle.

571

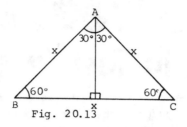

Fig. 20.13

Theorem: The area of an equilateral triangle equals $\frac{\sqrt{3}}{3}$ times the square of the altitude of the triangle.

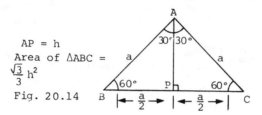

AP = h
Area of $\triangle ABC$ = $\frac{\sqrt{3}}{3} h^2$
Fig. 20.14

Theorem: A median drawn to a side of a triangle divides the triangle into two triangles of equal area.

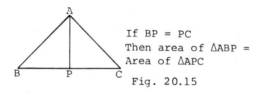

If BP = PC
Then area of $\triangle ABP$ =
Area of $\triangle APC$
Fig. 20.15

Theorem: The area of an isosceles triangle whose congruent sides have length 1, with included angle α is:

$$A = \tfrac{1}{2}1^2 \sin\alpha.$$

Area is also given by the formula

$$A = h^2 \tan \frac{\alpha}{2};$$

where h is the length of the altitude to the side opposite to the angle α.

REA's **Problem Solvers**